Environmental Interface Chemistry and Pollution Control

Environmental Interface Chemistry and Pollution Control

Editor

Ning Yuan

Basel • Beijing • Wuhan • Barcelona • Belgrade • Novi Sad • Cluj • Manchester

Editor
Ning Yuan
China University of
Mining and Technology
Beijing
China

Editorial Office
MDPI
St. Alban-Anlage 66
4052 Basel, Switzerland

This is a reprint of articles from the Special Issue published online in the open access journal *Sustainability* (ISSN 2071-1050) (available at: https://www.mdpi.com/journal/sustainability/special_issues/EnvironmentalInterface_PollutionControl).

For citation purposes, cite each article independently as indicated on the article page online and as indicated below:

Lastname, A.A.; Lastname, B.B. Article Title. *Journal Name* **Year**, *Volume Number*, Page Range.

ISBN 978-3-7258-0967-7 (Hbk)
ISBN 978-3-7258-0968-4 (PDF)
doi.org/10.3390/books978-3-7258-0968-4

© 2024 by the authors. Articles in this book are Open Access and distributed under the Creative Commons Attribution (CC BY) license. The book as a whole is distributed by MDPI under the terms and conditions of the Creative Commons Attribution-NonCommercial-NoDerivs (CC BY-NC-ND) license.

Contents

About the Editor . . . . **vii**

Ning Yuan
Resource Utilization of Solid Waste and Water Quality Evaluation
Reprinted from: *Sustainability* **2024**, *16*, 3189, doi:10.3390/su16083189 1

Tianxiang Chen, Ning Yuan, Shanhu Wang, Xinfei Hao, Xinling Zhang, Dongmin Wang and Xuan Yang
The Effect of Bottom Ash Ball-Milling Time on Properties of Controlled Low-Strength Material Using Multi-Component Coal-Based Solid Wastes
Reprinted from: *Sustainability* **2022**, *14*, 9949, doi:10.3390/su14169949 6

Yanbo Zhang, Ze Liu, Jixiang Wang, Conghao Shao, Jiaxing Li and Dongmin Wang
Study on the Hydration and Microstructure of B and B/Na Ion-Doped Natural Hydraulic Lime Composed with Silica Fume/Fly Ash
Reprinted from: *Sustainability* **2022**, *14*, 10484, doi:10.3390/su141710484 21

Haonan Cui, Haili Cheng, Tianyong Huang, Feihua Yang, Haoxiang Lan and Jvlun Li
Effect of Different Activators on Properties of Slag-Gold Tailings-Red Mud Ternary Composite
Reprinted from: *Sustainability* **2022**, *14*, 13573, doi:10.3390/su142013573 38

Zhaoyun Zhang, Chuang Xie, Zhaohu Sang and Dejun Li
Mechanical Properties and Microstructure of Alkali-Activated Soda Residue-Blast Furnace Slag Composite Binder
Reprinted from: *Sustainability* **2022**, *14*, 11751, doi:10.3390/su141811751 53

Zhaoyun Zhang, Chuang Xie, Zhaohu Sang and Dejun Li
Optimizing the Mechanical Performance and Microstructure of Alkali-Activated Soda Residue-Slag Composite Cementing Materials by Various Curing Methods
Reprinted from: *Sustainability* **2022**, *14*, 13661, doi:10.3390/su142013661 70

Alen Erjavec, Olivija Plohl, Lidija Fras Zemljič and Valh Volmajer Julija
Significant Fragmentation of Disposable Surgical Masks—Enormous Source for Problematic Micro/Nanoplastics Pollution in the Environment
Reprinted from: *Sustainability* **2022**, *14*, 12625, doi:10.3390/su141912625 84

Kai Zhang, Shunjie Wang, Shuyu Liu, Kunlun Liu, Jiayu Yan and Xuejia Li
Water Environment Quality Evaluation and Pollutant Source Analysis in Tuojiang River Basin, China
Reprinted from: *Sustainability* **2022**, *14*, 9219, doi:10.3390/su14159219 102

Wanli Su, Feisheng Feng, Ke Yang, Yong Zhou, Jiqiang Zhang and Jie Sun
Water Chemical Characteristics and Safety Assessment of Irrigation Water in the Northern Part of Hulunbeier City, Grassland Area in Eastern China
Reprinted from: *Sustainability* **2022**, *14*, 16068, doi:10.3390/su142316068 119

Linzhi Wang, Mingzhong Gao and Jiqiang Zhang
Effect of Continuous Loading Coupled with Wet–Dry Cycles on Strength Deterioration of Concrete
Reprinted from: *Sustainability* **2022**, *14*, 13407, doi:10.3390/su142013407 135

Baiping Li, Yunhai Cheng and Fenghui Li
Development and Constitutive Model of Fluid–Solid Coupling Similar Materials
Reprinted from: *Sustainability* **2023**, *15*, 3379, doi:10.3390/su15043379 150

About the Editor

Ning Yuan

Ning Yuan is an Associate Professor at the School of Chemical and Environmental Engineering, China University of Mining and Technology, Beijing, China. He earned his bachelor's degree in Polymer Material and Engineering from the Harbin Institute of Technology (2008), a master's degree in Inorganic Chemistry from the Chinese Academy of Sciences (2011), and a Ph.D. in Chemistry from the University of Tübingen, Germany (2015). He has worked at China University of Mining and Technology (Beijing) since 2016. His research interests include nanoporous materials and their application in the environment, architecture, energy, etc.

Editorial

Resource Utilization of Solid Waste and Water Quality Evaluation

Ning Yuan

School of Chemical and Environmental Engineering, China University of Mining and Technology, Beijing 100083, China; ning.yuan@cumtb.edu.cn

Citation: Yuan, N. Resource Utilization of Solid Waste and Water Quality Evaluation. *Sustainability* **2024**, *16*, 3189. https://doi.org/10.3390/su16083189

Received: 2 April 2024
Accepted: 9 April 2024
Published: 11 April 2024

Copyright: © 2024 by the author. Licensee MDPI, Basel, Switzerland. This article is an open access article distributed under the terms and conditions of the Creative Commons Attribution (CC BY) license (https://creativecommons.org/licenses/by/4.0/).

1. Introduction

The rapid development of industrialization and urbanization has inevitably resulted in the generation of innumerable solid wastes and water contamination. Solid waste refers to solid-state or semi-solid discarded trash that loses its original value and is discharged into the environment. According to various sources, solid waste can be divided into industrial, agricultural, and municipal solid waste. Solid waste not only occupies a great deal of space but may also cause pollution to air, soil, and water resources, and even affect human health. Therefore, it is necessary to implement the treatment and resource utilization of solid waste. The conventional treatment techniques include incineration, composting, landfill, and so forth. However, these treatment techniques would result in secondary pollution. Most of the solid waste contains recyclable components, which can be converted into various resources and transform trash to treasure [1].

With continuous industrial development, the resource utilization of industrial solid waste has aroused intense interest because of its diverse types, enormous quantity, and high resource utilization difficulty. Resource utilization can maximize the value of solid waste and improve social benefits. Some industrial residues have been applied in the fields of the construction industry, mine backfill, soil reclamation, and so forth. For example, crushed coal gangue and tailings can be utilized as aggregates, while fly ash and slag are commonly used as cementitious materials [2,3].

Except for large-scale utilization, the high value-added utilization of solid waste has been exploited due to its physicochemical properties [4]. The relevant applications include the extraction of rare metals and alumina, the fabrication of catalysts and molecular sieves, etc. [5,6]. Industrial residues with a high content of aluminosilicate have been employed for the fabrication of various types of porous materials, such as zeolites and mesoporous silicas [7,8]. Intriguingly, a type of crystalline nanoporous materials, metal–organic frameworks, have been recently synthesized from several categories of industrial solid waste [9,10]. The material utilization of industrial solid waste has become a hot topic, and beyond porous materials, many other novel materials have also been successfully obtained from industrial residues.

Human activities have not only resulted in the emission of diverse solid waste but also affected various water environments. Clean water is vital to individual health and public security. To access clean water resources, the evaluation of water quality and the analysis of water pollution are the prerequisites and foundations. Many efforts have been made in this area [11,12]. It is worth mentioning that solid waste-derived materials have been applied to purify the wastewater through physical, chemical, and biological methods. As a common bulk solid waste, coal fly ash has been widely investigated for the fabrication of adsorbents and catalysts, and the derived materials have exhibited excellent performance in wastewater treatment [13].

This Special Issue titled "Environmental Interface Chemistry and Pollution Control" in *Sustainability* aims to set up a forum for environmental pollution issues. The scope of this Special Issue covers research on environmental interface chemistry, pollution control, and environmental materials. In view of the published contributions, it is a collection of related research that mainly provides new insights into the resource utilization of solid waste and water quality evaluation.

2. Contributions

The collection of this Special Issue includes ten contributions, with themes mainly covering the resource utilization of solid waste and water quality evaluation. Among these contributions, six papers deal with the resource utilization and environmental influence of solid waste. The involved solid waste can be categorized into two types, namely, industrial solid waste and medical waste. The former includes coal fly ash, bottom ash, gasification slag, desulfurized gypsum, coal gangue, silica fume, blast furnace slag, gold tailings, red mud, soda residue, and so on. We will first present the contributions concerning the resource utilization of industrial solid waste.

Yuan et al. (Contribution 1) confirmed the feasibility of the production of controlled low-strength material (CLSM) by employing five types of coal-based solid wastes (coal fly ash, bottom ash, gasification slag, desulfurized gypsum, and coal gangue) and a small proportion of cement. The performances of the obtained fresh mortar (flowability, bleeding, setting time, and fresh density) and hardened mortar (compressive strength, porosity, absorption, and dry density) were carefully examined to evaluate whether they can meet the requirements of American Concrete Institute Committee 229. Furthermore, the effects of the mixing amount and ball-milling time of bottom ash on the properties of fresh and hardened CLSM specimens were studied. Also, the microstructures of CLSM specimens at 28-day curing age were investigated using the X-ray diffraction technique and scanning electron microscopy.

Liu et al. (Contribution 2) utilized fly ash and silica fume as supplementary cementitious materials (SCMs) to mix with the pre-synthesized natural hydraulic lime (NHL). The fabricated NHL was obtained by mixing limestone and diatomite and then doped with a small quantity of B_2O_3 or B_2O_3/Na_2CO_3 as a stabilizer. This work emphatically discussed the effects of ion doping (B and B/Na) and SCMs (fly ash and silica fume) on the compressive strength of NHL. The introduction of fly ash and silica fume can enhance the compressive strength of specimens, which was further improved by ion doping based on the coupling effect. The incorporation of silica fume remarkably boosted the early strength of the NHL specimens because of its high pozzolanic activity, whereas fly ash improved the late strength of the NHL specimens with consideration of its slower pozzolanic reaction degree. It is worth noting that the ion doping and introduction of SCMs can promote the hydration reaction and further enhance the compressive strength under the synergistic effect.

Cui et al. (Contribution 3) reported the blast furnace slag–gold tailings–red mud ternary composite cementitious material (SGRCM), which can mitigate environmental pollution and reduce production costs. Three types of bases (NaOH, KOH, and Na_2SiO_3) were adopted as activators for the ternary composite system. The compressive strength of alkali-activated SGRCM at a 28-day curing age can reach 43.7 MPa. According to the XRD, SEM, and EDS analyses, the main hydration products were C-A-S-H and ettringite, which were attributed to the enhancement of compressive strength. Moreover, the heavy metal ion leaching experiment results certified that the heavy metal ion can be consolidated in the SGRCM ternary composite cementitious material. The findings of this work demonstrated that Na_2SiO_3-activated blast furnace slag–gold tailings–red mud ternary composite cementitious material can replace cement to some extent.

Zhang et al. (Contribution 4) prepared a new alkali-activated composite binder based on soda residue and blast furnace slag and further investigated its mechanical properties and microstructure, which provided a solution to resolve the environmental pollution issue caused by alkali residues. The effects of Na_2O content, soda residue–blast furnace slag ratio, and water–binder ratio on the compressive strength at different curing ages (3, 7, 28, and 56 d) were explored. The optimal level of Na_2O content was 3.0%, at which the compressive strength was 27.8 MPa. As revealed in the XRD analysis, the main mineral compositions of the hydration products were C-(A)-S-H gel, ettringite, hydrocalumite, and calcium hydroxide. The experimental results disclosed that an appropriate Na_2O content can promote the structure density and enhance the strength, whereas an excessive Na_2O content would hinder strength development.

In subsequent work, Zhang et al. (Contribution 5) aimed to optimize the mechanical performance of alkali-activated soda residue–ground granulated blast furnace slag (SR-GGBS) cementing materials by regulating the curing methods, temperatures, and times. The findings showed that high-temperature curing can prominently enhance the early strength of the prepared SR-GGBS cementing materials. By comparing the compressive strengths and microstructural results, the optimal curing condition was determined to be 60 °C for 12 h. However, a further increase in temperature reduced the mechanical performance, and shrinkage cracks appeared in the test specimen. Therefore, in view of energy conservation, the room-temperature curing method can be considered in case of an undemanding requirement for early strength.

Besides industrial solid waste, an intriguing work from Slovenia on a kind of medical waste, disposable surgical masks (DSMs), is included in this Special Issue. Erjavec et al. (Contribution 6) focused on the problematic micro/nanoplastic pollution issue resulting from DSMs, with quantities dramatically amplified by the COVID-19 pandemic. In their work, DSMs from different sources in Slovenia were collected to make the waste samples more heterogeneous. During the leaching and degradation processes of DSMs, the potential water pollution triggered by micro/nanoplastics and other harmful components may emerge. The experimental findings revealed that DSMs contained a high content of chlorine elements, indicating the existence of halogenated organic compounds. Nevertheless, when DSMs were exposed to leaching, moisture, temperature, and UV radiation during a short period, a considerable amount of micro/nanoplastics were detected, which was the main problem. Finally, several recommended strategies were proposed, including stricter management, effective recycling, reuse, resource utilization, biodegradable alternatives, and ecotoxicological measures.

Two papers from China are presented on this topic of water quality evaluation. Zhang et al. (Contribution 7) evaluated the water environment quality and analyzed the pollution source of a river in southwest China (Tuojiang River) by selecting chemical oxygen demand, ammonia nitrogen, and total phosphorus as evaluation indicators. The results demonstrated that grey water footprint declined, suggesting an improvement in the water environment quality. Total phosphorus accounted for the largest percentage of the pollution sources of the grey water footprint. Farmland and stock breeding pollution were the dominant sources of the grey water footprint and contributed major influence factors for the water environment. This research suggested that the water environment quality in the Tuojiang River has been improving. On account of these findings, some constructive suggestions were proposed in this work. Feng and Yang et al. (Contribution 8) performed a water chemical characteristics and safety assessment of the Chenbarhu Banner coalfield in northeast China (Hulun Buir). The results disclosed that Na^+ and HCO_3^- were the dominant chemicals in the groundwater. The main chemical types of surface water, river water, and confined water were revealed, respectively. Further analysis illustrated that rock leaching and ion exchange were the dominant influencing factors of hydrochemical characteristics. This work showed that the groundwater quality was mainly polluted by breeding, industrial and agricultural pollution as well as domestic sewage instead of the local coal mine and coal chemical plants. The aforementioned studies provided significant references for the quality evaluation and pollution analysis of both surface water and groundwater.

Cement and concrete materials have shaped modern civilization. Two papers address this topic. Gao et al. (Contribution 9) investigated the effects of continuous loading and wet–dry cycles on the deterioration of concrete strength. The concrete strength was evaluated by uniaxial compression, and the strength deterioration mechanism was proposed by employing acoustic emission and nuclear magnetic resonance techniques. Li et al. (Contribution 10) developed a fluid–solid coupling similar material by adopting stone (5–20 mm) as aggregate and Portland cement (P.O 32.5) as a binder. The fluid–solid coupling similar materials possessed the following characteristics: loose structure, high porosity, and good permeability. The constructed material with a low uniaxial compressive strength of 0.394–0.528 MPa did not disintegrate in water. The influence of the water–cement ratio on the elastic modulus was studied, and the constitutive model was further established based on the Weibull distribution.

3. Conclusions

The contributions in this Special Issue will advance the research on the resource utilization of solid waste and water quality evaluation through new methods and strategies. This collection delves into the material utilization of several kinds of industrial residues, which realizes the generation of a series of materials, including controlled low-strength filling material and composite cementitious materials. Apart from industrial residues, the collection also discusses the potential micro/nanoplastic pollution from disposable surgical masks. Also, it presents the water environment quality of two different areas, viz., surface water in southwest China (Tuojiang River) and groundwater in northeast China (Hulun Buir). Moreover, two contributions are focused on cement-based materials, which exploit the deterioration of concrete under continuous loading and wet–dry cycles, and the excellent performance of a fluid–solid coupling similar material. The research presented in this collection will offer an important reference to further exploration in this realm. Future efforts can be made in the crossing research between the material utilization of solid waste and wastewater purification.

Funding: This work was financially supported by the National Key Research and Development Program of China (2019YFC1904304) and the Fundamental Research Funds for the Central Universities (2023ZKPYHH05).

Conflicts of Interest: The author declares no conflicts of interest.

List of Contributions:

1. Chen, T.; Yuan, N.; Wang, S.; Hao, X.; Zhang, X.; Wang, D.; Yang, X. The Effect of Bottom Ash Ball-Milling Time On Properties of Controlled Low-Strength Material Using Multi-Component Coal-Based Solid Wastes. *Sustainability* **2022**, *14*, 9949. https://doi.org/10.3390/su14169949.
2. Zhang, Y.; Liu, Z.; Wang, J.; Shao, C.; Li, J.; Wang, D. Study On the Hydration and Microstructure of B and B/Na Ion-Doped Natural Hydraulic Lime Composed with Silica Fume/Fly Ash. *Sustainability* **2022**, *14*, 10484. https://doi.org/10.3390/su141710484.
3. Cui, H.; Cheng, H.; Huang, T.; Yang, F.; Lan, H.; Li, J. Effect of Different Activators on Properties of Slag-Gold Tailings-Red Mud Ternary Composite. *Sustainability* **2022**, *14*, 13573. https://doi.org/10.3390/su142013573.
4. Zhang, Z.; Xie, C.; Sang, Z.; Li, D. Mechanical Properties and Microstructure of Alkali-Activated Soda Residue-Blast Furnace Slag Composite Binder. *Sustainability* **2022**, *14*, 11751. https://doi.org/10.3390/su141811751.
5. Zhang, Z.; Xie, C.; Sang, Z.; Li, D. Optimizing the Mechanical Performance and Microstructure of Alkali-Activated Soda Residue-Slag Composite Cementing Materials by Various Curing Methods. *Sustainability* **2022**, *14*, 13661. https://doi.org/10.3390/su142013661.
6. Erjavec, A.; Plohl, O.; Zemljič, L.F.; Valh, J.V. Significant Fragmentation of Disposable Surgical Masks—Enormous Source for Problematic Micro/Nanoplastics Pollution in the Environment. *Sustainability* **2022**, *14*, 12625. https://doi.org/10.3390/su141912625.
7. Zhang, K.; Wang, S.; Liu, S.; Liu, K.; Yan, J.; Li, X. Water Environment Quality Evaluation and Pollutant Source Analysis in Tuojiang River Basin, China. *Sustainability* **2022**, *14*, 9219. https://doi.org/10.3390/su14159219.
8. Su, W.; Feng, F.; Yang, K.; Zhou, Y.; Zhang, J.; Sun, J. Water Chemical Characteristics and Safety Assessment of Irrigation Water in the Northern Part of Hulunbeier City, Grassland Area in Eastern China. *Sustainability* **2022**, *14*, 16068. https://doi.org/10.3390/su142316068.
9. Wang, L.; Gao, M.; Zhang, J. Effect of Continuous Loading Coupled with Wet–Dry Cycles On Strength Deterioration of Concrete. *Sustainability* **2022**, *14*, 13407. https://doi.org/10.3390/su142013407.

10. Li, B.; Cheng, Y.; Li, F. Development and Constitutive Model of Fluid–Solid Coupling Similar Materials. *Sustainability* **2023**, *15*, 3379. https://doi.org/10.3390/su15043379.

References

1. Nallapaneni, M.K.; Hait, S.; Priya, A.; Bohra, V. From Trash to Treasure: Unlocking the Power of Resource Conservation, Recycling, and Waste Management Practices. *Sustainability* **2023**, *15*, 13863. [CrossRef]
2. Friol Guedes De Paiva, F.; Tamashiro, J.R.; Pereira Silva, L.H.; Kinoshita, A. Utilization of inorganic solid wastes in cementitious materials—A systematic literature review. *Constr. Build. Mater.* **2021**, *285*, 122833. [CrossRef]
3. Behera, S.K.; Mishra, D.P.; Singh, P.; Mishra, K.; Mandal, S.K.; Ghosh, C.N.; Kumar, R.; Mandal, P.K. Utilization of mill tailings, fly ash and slag as mine paste backfill material: Review and future perspective. *Constr. Build. Mater.* **2021**, *309*, 125120. [CrossRef]
4. Zhang, X.; Zhang, H.; Liang, Q.; Zhao, J.; Pan, D.; Ma, J. Resource utilization of solid waste in the field of phase change thermal energy storage. *J. Energy Storage* **2023**, *58*, 106362. [CrossRef]
5. Wang, C.; Xu, G.; Gu, X.; Gao, Y.; Zhao, P. High value-added applications of coal fly ash in the form of porous materials: A review. *Ceram. Int.* **2021**, *47*, 22302–22315. [CrossRef]
6. Mostafa Hosseini Asl, S.; Ghadi, A.; Sharifzadeh Baei, M.; Javadian, H.; Maghsudi, M.; Kazemian, H. Porous catalysts fabricated from coal fly ash as cost-effective alternatives for industrial applications: A review. *Fuel* **2018**, *217*, 320–342. [CrossRef]
7. Li, X.; Shao, J.; Zheng, J.; Bai, C.; Zhang, X.; Qiao, Y.; Colombo, P. Fabrication and Application of Porous Materials Made From Coal Gangue: A Review. *Int. J. Appl. Ceram. Technol.* **2023**, *20*, 2099–2124. [CrossRef]
8. Chen, X.; Zhang, P.; Wang, Y.; Peng, W.; Ren, Z.; Li, Y.; Chu, B.; Zhu, Q. Research progress on synthesis of zeolites from coal fly ash and environmental applications. *Front. Environ. Sci. Eng.* **2023**, *17*, 149. [CrossRef]
9. El-Sayed, E.M.; Yuan, D. Waste to MOFs: Sustainable linker, metal, and solvent sources for value-added MOF synthesis and applications. *Green Chem.* **2020**, *22*, 414–482. [CrossRef]
10. Shanmugam, M.; Chuaicham, C.; Augustin, A.; Sasaki, K.; Sagayaraj, P.J.J.; Sekar, K. Upcycling Hazardous Metals and Pet Waste-Derived Metal-Organic Frameworks: A Review on Recent Progresses and Prospects. *New J. Chem.* **2022**, *46*, 15776–15794. [CrossRef]
11. Sharma, R.; Kumar, R.; Sharma, D.K.; Sarkar, M.; Mishra, B.K.; Puri, V.; Priyadarshini, I.; Thong, P.H.; Ngo, P.T.T.; Nhu, V. Water pollution examination through quality analysis of different rivers: A case study in India. *Environ. Dev. Sustain.* **2022**, *24*, 7471–7492. [CrossRef]
12. Ighalo, J.O.; Adeniyi, A.G.; Adeniran, J.A.; Ogunniyi, S. A systematic literature analysis of the nature and regional distribution of water pollution sources in Nigeria. *J. Clean. Prod.* **2021**, *283*, 124566. [CrossRef]
13. Hosseini Asl, S.M.; Javadian, H.; Khavarpour, M.; Belviso, C.; Taghavi, M.; Maghsudi, M. Porous Adsorbents Derived From Coal Fly Ash as Cost-Effective and Environmentally-Friendly Sources of Aluminosilicate for Sequestration of Aqueous and Gaseous Pollutants: A Review. *J. Clean. Prod.* **2019**, *208*, 1131–1147. [CrossRef]

Disclaimer/Publisher's Note: The statements, opinions and data contained in all publications are solely those of the individual author(s) and contributor(s) and not of MDPI and/or the editor(s). MDPI and/or the editor(s) disclaim responsibility for any injury to people or property resulting from any ideas, methods, instructions or products referred to in the content.

Article

The Effect of Bottom Ash Ball-Milling Time on Properties of Controlled Low-Strength Material Using Multi-Component Coal-Based Solid Wastes

Tianxiang Chen [1], Ning Yuan [1,*], Shanhu Wang [1], Xinfei Hao [2], Xinling Zhang [1], Dongmin Wang [1] and Xuan Yang [1]

[1] School of Chemical and Environmental Engineering, China University of Mining and Technology, Beijing 100083, China
[2] Huayang New Material Technology Group Co., Ltd., Yangquan 045000, China
* Correspondence: ning.yuan@cumtb.edu.cn

Abstract: As the conventional disposal method for industrial by-products and wastes, landfills can cause environmental pollution and huge economic costs. However, some secondary materials can be effectively used to develop novel underground filling materials. Controlled low-strength material (CLSM) is a highly flowable, controllable, and low-strength filling material. The rational use of coal industry by-products to prepare CLSM is significant in reducing environmental pollution and value-added disposal of solid waste. In this work, five different by-products of the coal industry (bottom ash (BA), fly ash, desulfurized gypsum, gasification slag, and coal gangue) and cement were used as mixtures to prepare multi-component coal industry solid waste-based CLSM. The microstructure and phase composition of the obtained samples were analyzed by scanning electron microscopy and X-ray diffraction. In addition, the particle size/fineness of samples was also measured. The changes in fresh and hardened properties of CLSM were studied using BA after ball milling for 20 min (BAI group) and 45 min (BAII group) that replaced fly ash with four mass ratios (10 wt%, 30 wt%, 50 wt%, and 70 wt%). The results showed that the CLSM mixtures satisfied the limits and requirements of the American Concrete Institute Committee 229 for CLSM. Improving the mass ratio of BA to fly ash and the ball-milling time of the BA significantly reduced the flowability and the bleeding of the CLSM; the flowability was still in the high flowability category, the lowest bleeding BAI70 (i.e., the content of BA in the BAI group was 70 wt%) and BAII70 (i.e., the content of BA in the BAII group was 70 wt%) decreased by 48% and 64%, respectively. Furthermore, the 3 d compressive strengths of BAI70 and BAII70 were increased by 48% and 93%, respectively, compared with the group without BA, which was significantly favorable, whereas the 28 d compressive strength did not change significantly. Moreover, the removability modulus of CLSM was calculated, which was greater than 1, indicating that CLSM was suitable for structural backfilling that requires a certain strength. This study provides a basis for the large-scale utilization of coal industry solid waste in the construction industry and underground coal mine filling.

Keywords: controlled low-strength material (CLSM); coal industry by-products; bottom ash; bleeding; compressive strength

Citation: Chen, T.; Yuan, N.; Wang, S.; Hao, X.; Zhang, X.; Wang, D.; Yang, X. The Effect of Bottom Ash Ball-Milling Time on Properties of Controlled Low-Strength Material Using Multi-Component Coal-Based Solid Wastes. *Sustainability* **2022**, *14*, 9949. https://doi.org/10.3390/su14169949

Academic Editor: Dimitrios Komilis

Received: 18 June 2022
Accepted: 9 August 2022
Published: 11 August 2022

Publisher's Note: MDPI stays neutral with regard to jurisdictional claims in published maps and institutional affiliations.

Copyright: © 2022 by the authors. Licensee MDPI, Basel, Switzerland. This article is an open access article distributed under the terms and conditions of the Creative Commons Attribution (CC BY) license (https://creativecommons.org/licenses/by/4.0/).

1. Introduction

According to the American Concrete Institute (ACI) Committee 229, controlled low-strength material (CLSM) is a self-compacting and self-leveling cementitious material [1], usually consisting of water, Portland cement, fly ash, and fine aggregate. CLSM can be used in areas that are difficult to access and do not require compression because of its high flowability. The compressive strength of CLSM is designable and determined by whether or not excavation is required in the future. When manual excavation is required, the recommended compressive strength is less than 0.3 MPa, and when mechanical excavation is required, the recommended compressive strength is between 0.7 and 1.4 MPa. If excavation is not necessary, the recommended compressive strength is less than 8.3 MPa [2,3]. Due to its low strength requirements, CLSM can be perfectly combined with various wastes and by-products obtained locally, thereby reducing costs and improving filling quality [4]. Therefore, CLSM is widely used in geotechnical engineering, such as coal mine fills, pipeline bedding, void fills, etc. Over the years, CLSM has been tested with numerous solid wastes to replace fine aggregates. These include quarry waste [5], treated oil sands waste [6], coal-fired products [7], desulfurization slag [8], etc. To further reduce material costs, some scholars have considered using ground granulated blast-furnace slag [9], dust from asphalt plants [10], waste glass powder [11], calcium carbide slag [12], wastewater sludge [13], etc. to substitute for cement as cementitious materials when producing CLSM. In addition, additives such as foam particles, air-entraining agents, and water-reducing agents can be added to obtain freshly mixed CLSM with high workability.

With the exploitation, processing, and utilization of coal, China produces a great deal of coal-based wastes, such as bottom ash (BA), coal gangue, gasification slag, fly ash, and desulfurized gypsum [14,15]. However, these coal-based solid wastes are generated, stockpiled, and not effectively utilized, which causes serious ecological damage and environmental pollution [16,17]. In order to address this problem, some researchers have developed a series of coal-based solid wastes resource utilization approaches [18–20], including the preparation of functional materials from wastes, coal gangue power generation, agricultural application, valuable element recovery, and so forth. However, as China's economic development gradually enters the new normal, the demand for traditional building materials such as cement and concrete has decreased in the building materials industry [21]. This makes the comprehensive utilization of coal-based solid waste face severe challenges. If CLSM prepared from multi-component coal-based solid wastes can be applied to underground filling on a large scale, it can not only reduce the threat to the environment but also make reasonable use of solid wastes, resulting in huge economic benefits.

Using solid wastes and improving their comprehensive utilization rate is important in China's current ecological and environmental protection work. BA is the solid waste discharged from the bottom of a boiler after coal combustion in a power plant. The BA is mainly irregular with smooth edges and corners, mainly composed of SiO_2, Al_2O_3, CaO, and Fe_2O_3, accounting for about 90%, and belongs to silicon-alumina materials [22]. BA has good pozzolanic activity and can be used as a concrete admixture after being optimized by fine grinding or chemical modification [23,24]. For example, Pakawat et al. studied the variation of notch size particles on BA pozzolanic activity, and the results showed that the compressive strength values of mortar prepared with BA were 5% and 14% higher than that of mortar using 100% Portland cement at 28 d and 60 d of curing ages, respectively. The high pozzolanic activity of BA exhibited great potential as a cement substitute [25]. Additionally, Belén et al. used ground ash as a cement additive and compared its performance with that of limestone as an additive. The results showed that the compressive strength of cement with 40 wt% BA was higher than that of standard cement and limestone as additives. Furthermore, cement with 20 wt% BA can save 9% clinker against commercial cement (CEM II/AL 42.5 R), and cement with 40 wt% BA can save 14% clinker against commercial cement (CEM II/BL 32.5 N) [26]. Although BA has been utilized in cement and concrete, the utilization rate of BA and other wastes is not enough in general and needs to be improved.

In addition, researchers also selected other coal-based solid wastes for utilization [27]. Kaliyavaradhan et al. reviewed the effects of solid wastes such as circulating fluidized bed combustion ash, ponded ash, slag, and BA on the strength of CLSM and suggested the mixing ratio of the mixed waste in CLSM [28]. Ling et al. outlined the research development and practical application of CLSM in trench backfilling and found through 115 pieces of literature from various countries around the world that the materials used to produce CLSM vary from country to country. This will affect the nature of CLSM production and have significant implications for practical applications in the field [29]. Specifically, Zhang et al. used fly ash and coal gangue as raw materials to study the feasibility of filling. When the mass ratio of coal gangue:fly ash:cement was 14:5:1, the filling body not only had good fluidity but also could meet the requirements of compressive strength and dehydration [30]. With the development of CLSM, more kinds of solid waste are required to be added to the preparation of CLSM. For example, Yang et al. prepared composites using desulfurized gypsum, low-calcium fly ash, slag, and cement as cementitious materials and total tailings as aggregates and conducted a filling test. It is noted that when the fly ash content is high, the prepared composite materials can meet the strength requirements of mine filling [31]. The use of a wider variety of solid wastes is an effective solution to minimize the environmental problems associated with the disposal of these wastes.

At present, although CLSM based on the coal industry by-products has been investigated extensively, the research on the preparation of CLSM aiming at fully utilizing the five kinds of coal-based solid wastes remains scarce. In this work, for the first time, we mixed these five coal-based solid wastes with cement for the purpose of underground filling. In addition, in order to further reduce the bleeding of CLSM and improve the early strength of CLSM, the BA was ground. By changing the grinding time of the BA, we studied the effect of different ball-milling of BA on the flowability, bleeding, compressive strength (3 d, 7 d, and 28 d), fresh density, dry density, setting time, porosity, absorption, and microstructure of CLSM in detail. The preparation of this CLSM can not only reduce the environmental protection tax levied by some enterprises for stacking coal-based solid waste and reduce the burden on enterprises but also solve the current ecological and environmental problems caused by the inability to properly handle coal-based solid wastes. Specifically, in the obtained CLSM, BA, desulfurized gypsum, fly ash, and cement were employed as cementitious materials, and coal gangue and gasification slag were utilized as aggregates. BA with ball-milling times of 20 min and 45 min was used to replace fly ash in different mass ratios (i.e., 10 wt%, 30 wt%, 50 wt%, 70 wt%), and the changes in fresh and hardened properties of CLSM were investigated. The crystallinity and composition of the hydration products were characterized by X-ray diffraction (XRD), and the microstructure of CLSM was characterized by scanning electron microscopy (SEM).

2. Experimental Programs

2.1. Materials

The cement employed in the experiment is the benchmark cement PI 42.5 Portland cement. The chemical compositions of all coal-based solid wastes and cement are shown in Table 1, and their XRD characterizations are provided in Figure 1. The admixture used in the experiment is a powdered polycarboxylate superplasticizer.

Table 1. Chemical compositions of coal-based solid wastes and cement (wt%).

Compound	Cement	Bottom Ash	Fly Ash	Desulfurized Gypsum	Gasification Slag	Coal Gangue
SiO_2	20.72	56.37	52.95	2.62	48.07	48.46
Al_2O_3	4.62	26.71	27.55	0.58	16.37	24.13
CaO	62.18	3.41	4.94	28.77	8.95	0.10
Fe_2O_3	3.26	6.62	6.31	0.43	8.84	9.44
MgO	3.15	1.20	1.92	2.46	1.91	0.47
Na_2O	0.52	1.08	1.52	0.25	1.77	0.25
K_2O	0.34	1.58	1.85	0.12	1.48	1.99
TiO_2	–	1.04	1.28	0.03	0.90	0.86
SO_3	2.72	0.47	1.03	40.17	0.61	0.09
f-CaO	0.72	–	–	–	–	–
Cl^-	0.012	–	–	–	–	–
Loss	1.84	1.09	0.19	24.50	10.30	14.03

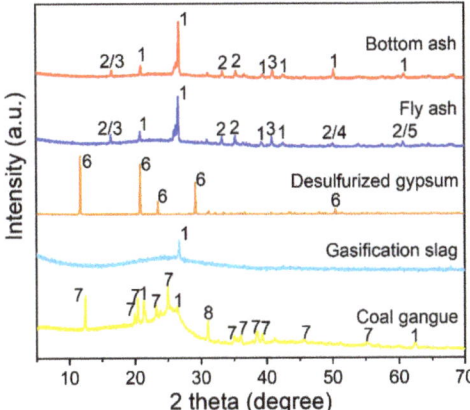

Figure 1. XRD patterns of bottom ash, fly ash, desulfurized gypsum, gasification slag, and coal gangue. Components: 1-SiO_2, 2-$3Al_2O_3 \cdot 2SiO_2$, 3-$Al_2[SiO_4]O$, 4-Fe_2O_3, 5-CaO, 6-$CaSO_4(H_2O)_2$, 7-$Al_2Si_2O_5(OH)_4$, 8-FeS_2.

BA used in the experiment is from Yuanyang Lake Power Plant in Ningxia, China. It can be seen in Figure 2 that the particle size of the BA changes after ball milling. After 20 min of ball milling, the particle size of the BA basically reaches 100 µm or less. When the ball milling time is increased to 45 min, the coarse BA is further crushed and refined, and the d_{50} decreases from 26.3 µm to 13.2 µm. It is close to fly ash particle size, but it is still larger than the latter. Further increasing the ball milling time, the proportion of fine particles will continue to increase. However, at the same time, the cost of the ball-milling process will increase exponentially. After comprehensive consideration, the ball-milling times of 20 min and 45 min are used for the pretreatment of BA in the experiment. Figure 3a presents the microstructure and morphology of the BA, which is mainly composed of irregular block particles with large differences in particle size, and the block particles display a porous surface.

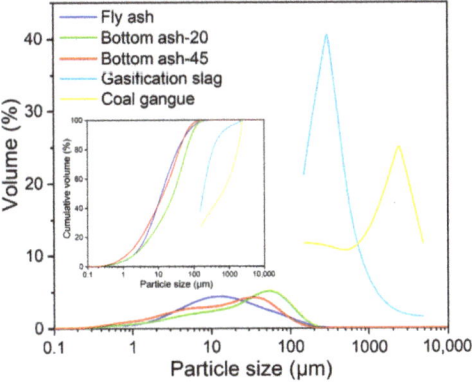

Figure 2. Particle size distributions of bottom ash at different ball-milling times (20 min, 45 min), fly ash, gasification slag, and coal gangue.

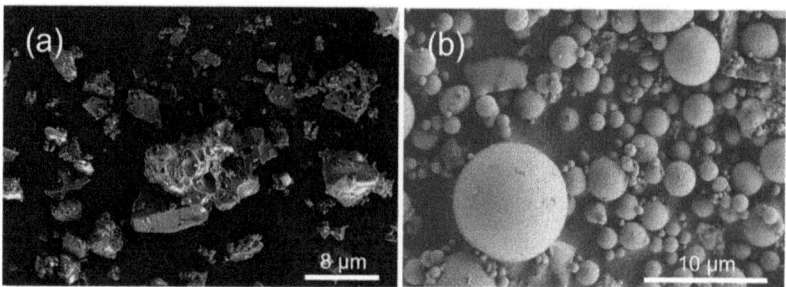

Figure 3. SEM analysis of (**a**) bottom ash and (**b**) fly ash.

Fly ash is the fine ash collected during the purification of flue gas in the second phase at the Yuanyang Lake Power Plant in Ningxia, China. According to Figure 2, the characteristic parameter of fly ash particle size d_{10}, d_{50}, and d_{90} are 2.51 μm, 13.18 μm, and 69.18 μm, respectively. Figure 3b presents the microstructure and morphology of fly ash, indicative of spherical particles.

The desulfurization gypsum is from the Yuanyang Lake Power Plant in Ningxia, China. It can be seen in Figure 1 that there is no other mineral phase except dihydrate gypsum, and the phase composition is almost single. The gasification slag used in the experiment is the coarse gasification slag from the Ningxia Coal Mine. The sieving results of the gasification slag are shown in Figure 2. The fineness module of the gasification slag is 1.73, and the particles between 0.15–0.6 mm account for 75%, which can be classified as fine sand. The coal gangue used in the experiment is the original gangue from the Renjiazhuang Coal Mine in Ningxia. It is crushed into gangue particles with a maximum particle size of less than 4.75 mm by the secondary jaw crusher. The crushed coal gangue is screened and analyzed, as shown in Figure 2, and it is found that the content of particles larger than 4.75 mm in the crushed coal gangue is only 11.7%. After further sieving analysis of particles smaller than 4.75 mm, it is found that in the crushed coal gangue smaller than 4.75 mm and the content of particles smaller than 0.15 mm reaches 15.3%. The fineness module of coal gangue is 2.83, which can be recognized as medium sand.

2.2. Mixture Proportions

The CLSM mixture proportions are shown in Table 2. Fly ash and two types of BA are used as the main cementitious materials to prepare CLSM mixtures, and the CLSM mixtures are divided into two groups: BAI group (BA with ball milling time of 20 min) and BAII group (BA with ball milling time of 45 min), to investigate the effect of the two types of BA on the CLSM mixture's performance. According to the recommendation of ACI Committee 229 [1], binder–aggregate ratios are lightly adjusted. The proportion of each aggregate and each cementitious material is determined according to previous research in our laboratory. Firstly, 60 wt% coal gangue (total solid mass of mixture) and 15 wt% gasification slag (total solid mass of mixture) are determined as fine aggregates, the ratio of aggregate to cementitious material is kept at 3:1, and the amounts of cement and desulfurized gypsum are 5 wt% and 1 wt%, respectively. Afterward, 10 wt%, 30 wt%, 50 wt%, and 70 wt% of the fly ash are replaced with BA according to the mass of the fly ash. The mixtures are named Blank, BAI10, BAI30, BAI50, BAI70, BAII10, BAII30, BAII50, and BAII70 according to the type and weight of the added BA, as shown in Table 2.

2.3. CLSM Preparation and Testing Procedure

All materials used in the experiments were naturally dried before use. Next, CLSM is prepared according to the mixture proportions (Table 2) and the mixing program. The ingredients were mixed for 1 min without adding water in a cement mortar mixer (Shanghai INESA Scientific Instrument, JJ-15 type) to ensure uniform distribution. Thereafter, the mixed water was divided in half. The first half of the mixed water and the polycarboxylate water-reducing agent were gradually added to the mixture while continuing to mix and stir for 1 min. Then, the second half of the mixed water was added, mixed, and stirred for 1 min. After all the mixed water was added, the mixture was allowed to stand for 1 min and then mixed for 2 min. The fresh mortar specimen was introduced into a mold of 70.7 mm × 70.7 mm × 70.7 mm. The mold was removed after 36 h, and the mortar specimen was cured in a standard curing room (SHBY-40B type, Cangzhou Huaxi, temperature 20 ± 1 °C, humidity above 95%) for 3 d, 7 d, and 28 d. All performance test standards are shown in Table 3.

Table 2. Experimental scheme of CLSM.

Mixture	Aggregate (kg m^{-3})		Cementitious Materials (kg m^{-3})				Water (kg m^{-3})	Water-Reducing Agent (kg m^{-3})
	Coal Gangue	Gasification Slag	Cement	Fly Ash	Bottom Ash	Desulfurized Gypsum		
Blank	960	240	80	304	0	16	280	0.8
BAI10	960	240	80	273.6	30.4	16	280	0.8
BAI30	960	240	80	212.8	91.2	16	280	0.8
BAI50	960	240	80	152	152	16	280	0.8
BAI70	960	240	80	91.2	212.8	16	280	0.8
BAII10	960	240	80	273.6	30.4	16	280	0.8
BAII30	960	240	80	212.8	91.2	16	280	0.8
BAII50	960	240	80	152	152	16	280	0.8
BAII70	960	240	80	91.2	212.8	16	280	0.8

Table 3. Standards for the testing methods.

Test Procedure	Standard	References
Flowability	ASTM D6103-17	[32]
Bleeding and fresh density	GB/T 50080-2016	[33]
Compressive strength	GB/T 50081-2019	[34]
Setting time	GB/T 1346-2011	[35]
Absorption, porosity, and dry density	ASTM D6023-16	[36]

ASTM-American Society for Testing and Materials; GB-China National Standard.

2.4. Microstructure Testing

X-ray diffraction (XRD) was acquired on a Rigaku Mini Flex 600 powder diffractometer using a Cu-Kα target (λ = 1.5406 Å) with a test voltage of 40 kV and a test current of 15 mA. The obtained diffraction pattern was compared with the standard pattern, and the composition of the sample was determined according to the position of the diffraction peak. The particle size/fineness analysis was performed using a Malvern Mastersizer 2000 laser particle size analyzer. The morphology and hydration products of the solid wastes were scanned using a cold field emission scanning electron microscope (ZEISS Gemini 300) at 25 °C with a cold field emission electron source, accelerating voltage of 0.5–30 kV and accelerating current of 10 mA.

3. Results and Discussion

3.1. Flowability

The flowability results for tested CLSM mixtures are shown in Figure 4. The flowability for all mixtures is between 193–280 mm. According to ACI Committee 299 [1], except for BAII70, all CLSM mixtures fall into the high flowability category. Compared with the Blank group, the flowability declines with the rise in both BA contents. This is because the shape of the BA particles is irregular and cannot function as a "ball bearing", which increases the internal friction between the particles [37]. At the same time, the pore volume of BA is larger than that of fly ash, and the rise in BA will cause more water to be absorbed into the inner pores of BA and reduce the flowability [38]. Furthermore, the flowability of the BAII group is slightly lower than that of the BAI group, which is due to the fact that with the rise in ball milling time, the BA particles become finer, which rises the pore volume and specific surface area, resulting in lower flowability [39,40]. This is also confirmed in the bleeding experiment results.

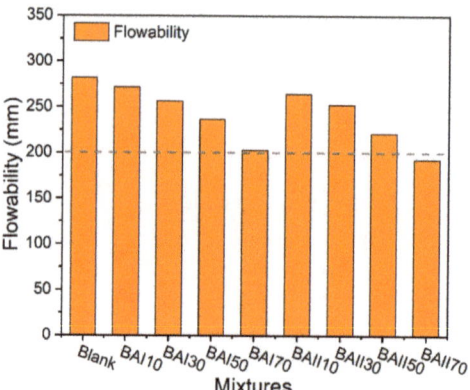

Figure 4. Flowability of various CLSM mixtures.

3.2. Bleeding

All bleeding results for the tested CLSM mixtures are shown in Figure 5. It can be confirmed that the CLSM does not exceed the 5% requirement recommended by ACI Committee 229 [1]. The bleeding value of the CLSM mixture declines with the rise in BA content. The bleeding values of BAI10, BAI30, BAI50, and BAI70 are 2.70%, 2.15%, 1.99%, and 1.53%, respectively, whereas the bleeding values of CLSM mixtures of BAII10, BAII30, BAII50, and BAII70 are 2.17%, 1.65%, 1.46%, and 1.05%, respectively, which are lower than 2.92% in the Blank mixture. Especially, BAI70 and BAII70 present a significant decrease of 48% and 64% in bleeding, respectively. There are two reasons for the reduction in bleeding. Firstly, the BA has a large particle size and rough and porous surface, which entitles the BA to a higher water-holding capacity [41], and secondly, the addition of BA reduces the morphological effect and micro-aggregate effect of fly ash. Because the shape of fly ash particles is suitable and the surface is smooth and dense, CLSM can obtain good water reduction. Simultaneously, gaps are formed between particles to prevent agglomeration between particles. However, this results in more permeable water from the CLSM, thus higher bleeding [42].

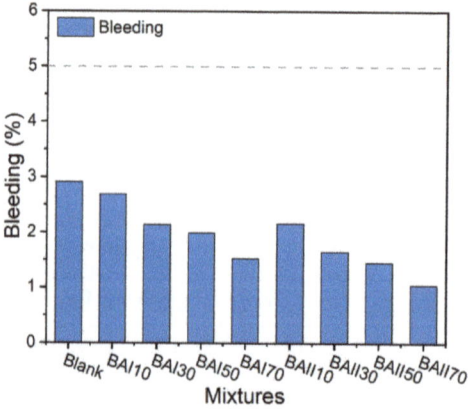

Figure 5. Bleeding of various CLSM mixtures.

In addition, when the mass ratio remains unchanged, the bleeding of the mixture of the BAII group is lower than that of the mixture of the BAI group. With the increase in BA ball milling time, finer BA particles result in a larger specific surface area and pore volume, an increase in water holding capacity, and a reduction of water exudation [43]. As a consequence, it can be concluded that the coarse and porous BA can absorb more water during the CLSM mixing process, which will reduce the exudation of water in the CLSM mixture. This phenomenon has also been observed in previous research [7,44,45].

3.3. Density

The density of the fresh and hardened CLSM samples at 28 d is determined, as displayed in Figure 6. The fresh density of CLSM ranges from 1940 kg m^{-3} to 2023 kg m^{-3}. Compared with the blank group, the fresh density of CLSM with BA is increased or decreased but still within the normal CLSM range (i.e., 1842 kg m^{-3}–2323 kg m^{-3}) reported by the ACI Committee 229 [1]. The BA with a long ball-milling time adsorbs more water. Therefore, with the same mass ratio, the fresh density of the BAII group is higher than that of the BAI group. This is also in line with the measurement results of bleeding.

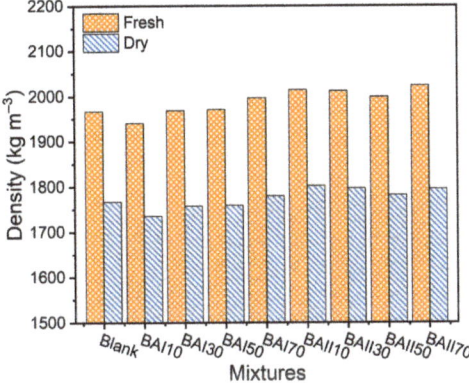

Figure 6. Fresh density and dry density test results of CLSM mixtures.

It can be seen that for all tested CLSM mixtures, the dry density is lower than the fresh density due to the loss of water in the CLSM samples [46]. Moreover, the dry density is consistent with the changing trend of the fresh density. The dry density of CLSM is between 1735 kg m^{-3} and 1801 kg m^{-3}, which is basically in line with CLSM requirements (i.e., 1762 kg m^{-3}–1890 kg m^{-3}) [1].

3.4. Compressive Strength

Figure 7 exhibits the compressive strength (3 d, 7 d, and 28 d) and removability modulus (RE) of CLSM. It can be seen that the compressive strength does not exceed the upper limit (8.3 MPa) recommended by the ACI Committee 229. When the amount of BA is the same, the compressive strength (3 d) of the CLSM mixture BAII group is higher than that of the BAI group (Figure 7a). Compared with Blank, as the amount of BA rises, the BAI group increases by 4%, 7%, 16%, and 48%, respectively, and the BAII group increases by 8%, 16%, 60%, and 93%, respectively. This can be explained as follows. The water absorption of BA is higher than that of fly ash, and the BA reduces the free water in the freshly mixed CLSM, which is helpful for the improvement of early strength [47]. Meanwhile, the high content of SiO$_2$ in the BA leads to the initial production of many calcium silicate hydrates (C-S-H) [38,48,49], resulting in a high early compressive strength of the CLSM prepared from the BA.

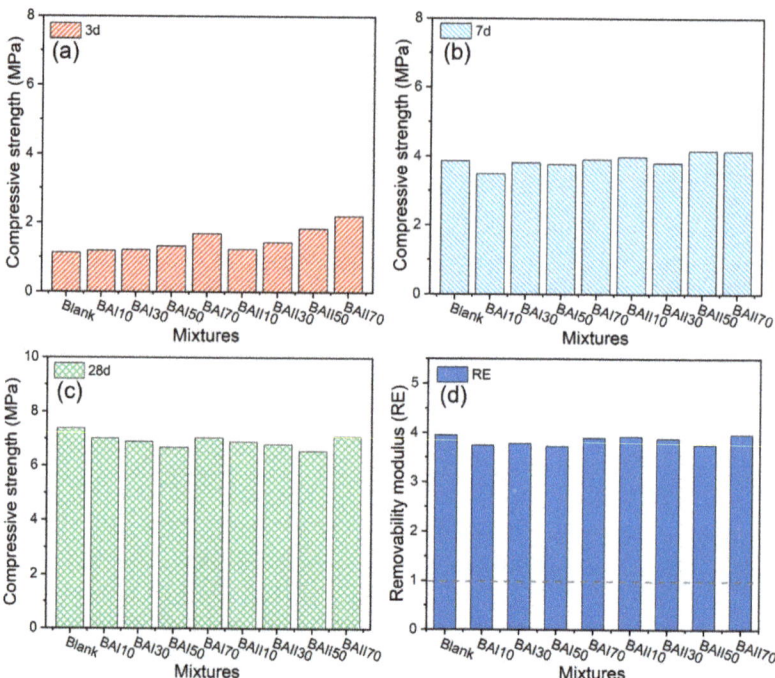

Figure 7. Compressive strength of CLSM mixtures at (**a**) 3 d, (**b**) 7 d, (**c**) 28 d, and (**d**) calculated results of removability modulus (RE).

With the rise in curing time, the development speed of strength (7 d) of CLSM with BA decreases and will tend to be consistent with the Blank group, but the trend of BAII > BAI > Blank can still be observed, as shown in Figure 7b. This is because the hydration reaction continues, the structure of the slurry becomes denser, and the non-homogeneity [7] of the CLSM containing BA begins to manifest, resulting in a decline in the speed of strength development. Compared with the BAI group, the BAII group possesses higher strength at 3 d and 7 d because the BAII group has an enhanced specific surface area and pozzolanic activity [26] after a long time of ball milling, and, thus, the formation of C-S-H gel is promoted. The aforementioned changes in compressive strength are consistent with previous studies [50]. In particular, in contrast to the blank group, the compressive strength of the BAI group exhibits a trend of first decreasing and then increasing, and the same trend can be observed in the BAII group. The lowest point of compressive strength in the BAI group appears when the BA content is 10% (BAI10), whereas the lowest value of compressive strength in the BAII group emerges with a BA content of 30% (BAII30). The occurrence of the lowest point moves backward. This is because the increase in the BA ball milling time (i.e., the decrease in bottom ash particle size) makes the non-homogeneity of CLSM only start to appear obvious after the content of BA is increased.

Finally, as shown in Figure 7c, it is observed that the compressive strength of CLSM at 28 d showed a change of Blank > BAI > BAII, and with the rise in BA, the 28 d compressive strength showed a trend of first declining and then rising. The reason for this change is that the pozzolanic activity of BA is not as high as that of fly ash, which contributes less to the compressive strength of CLSM in the later stage of curing, and the non-homogeneity of CLSM has a negative impact on the compressive strength. Therefore, the compressive strength of CLSM with BA is lower. It is worth noting that the mixtures containing 70 wt% BA have a higher 28 d compressive strength, which may be due to the large rise in BA content will lead to the decrease in CLSM non-homogeneity and the increase in CLSM compactness and, thus, the enhancement of compressive strength.

Figure 7d shows the removability modulus (RE) values for all tested CLSM mixtures. RE can be calculated according to Equation (1), based on the 28 d compressive strength (C) (kPa) and dry density (W) (kg m^{-3}). The RE value is used to evaluate the excavatability of CLSM mixtures. Structural filling applications require the CLSM to have sufficient load-carrying capacity. However, for projects that require later excavation, keeping the low strength is a major goal. Some early acceptable mixtures continue to build in strength over time, making future excavations difficult. BA has little effect on the removability modulus, with RE above 1 for all samples. Therefore, the operation of manually excavating these CLSM mixtures is difficult, and it is suitable for structural fills where certain strength is required [6,51].

$$RE = \frac{0.619 \times W^{1.5} \times C^{0.5}}{10^6} \quad (1)$$

3.5. Setting Time

For applications such as backfills, void filling, and structural fills, it is necessary to measure the setting time required by the CLSM to allow people or items to move over the CLSM surface. The maximum acceptable limit for the initial setting time of the CLSM mixture is 36 h. As shown in Figure 8, the initial setting time of our prepared CLSM mixture is between 4.35 and 7.48 h, which are all lower than the general CLSM requirements [52]. It is found that the increase on BA content accelerates the initial setting time of CLSM, which is the same as the results of previous studies [53,54]. The ball-milling time of BA exerts no significant effect on the initial setting time. Furthermore, the final setting time also declines significantly with the rise in BA content. The final setting time of BAI70 and BAII70 is reduced by 37% and 36%, respectively. The main reason for this phenomenon is that aluminum, silicon, and calcium can be leached more easily from BA than from fly ash [55], and C-S-H gel can also be formed more quickly, while the C-S-H gel plays a key role in the solidification behavior of CLSM. Therefore, this may be the reason why the initial and final setting time of CLSM decline with the rise in BA content.

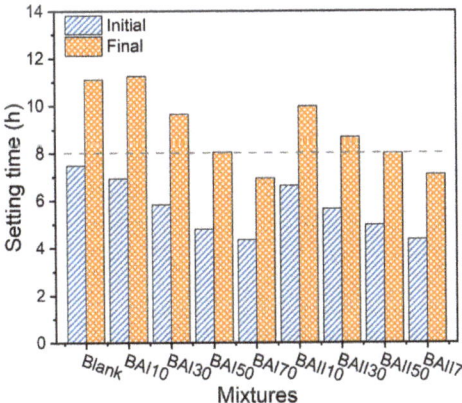

Figure 8. Initial and final setting time of prepared CLSM mixtures.

3.6. Absorption and Porosity

Absorption and porosity were tested at 28 d, and the data are depicted in Figure 9. The absorptivity of the CLSM mixtures is in the range of 13.5–14.9%, and the porosity is in the range of 11.9–13.0%, both of which increase with the rise in BA content. This is due to the fact that the pozzolanic activity of the BA is less than that of fly ash, resulting in less C-S-H gel produced, which cannot fill the macropores in the CLSM, thus forming an open microstructure [46].

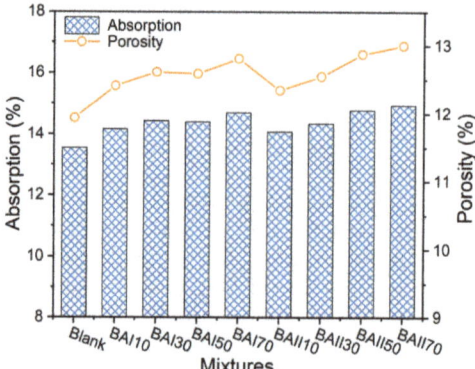

Figure 9. Absorption and porosity test results of prepared CLSM mixtures.

3.7. Microstructure

Figure 10a–f are the SEM images of CLSM at 28 d, showing the micro-interface properties of Blank, BAI30, BAI70, BAII30, and BAII70. It can be observed that unreacted fly ash microspheres are embedded in the gel phase and aggregates since fly ash particles of different sizes are spherical and easy to identify [46]. The existence of unreacted fly ash particles is caused by the excessive silica content in the system, which is further proved by the XRD results.

The presence of a large number of pores can be observed in Figure 10c–f. On the one hand, this is because the BA is a porous microstructure, which gives rise to high water absorption. The higher initial water content will lead to more voids during the hardening stage. On the other hand, this is due to the fact that the pozzolanic activity of BA is not as good as that of fly ash, and the hydration process produces less C-S-H gel, which cannot fill the macropores in CLSM, and finally forms an open microstructure with high porosity [37]. Compared with BAI70 and BAII70, the structures of BAI30 and BAII30 are slightly denser. This is also consistent with the test results of absorption and porosity.

Figure 11 presents the XRD patterns of the CLSM curing age of 28 d, in which diffraction peaks assigned to quartz, ettringite, C-S-H, and dehydrated gypsum can be observed [45,52]. Weak peaks (i.e., low-intensity peaks) for C-S-H and ettringite can be discovered in all CLSM samples. The hydration reaction produces ettringite and C-S-H, while quartz and gypsum crystals still exist as minerals themselves, presumably from fly ash, BA, and desulfurized gypsum. The incompletely reacted cementitious materials (such as fly ash) can also be found in the test blocks. These findings can also be demonstrated by SEM results. The quartz peak at 26.6° is dominant in the CLSM mixture. The quartz diffraction peaks of BAII70 and BAI70 are significantly stronger than that of Blank, BAII30, and BAI30. This is because the SiO_2 content in the chemical composition of the BA is higher than that in the fly ash, and the increase in the content of the BA will also lead to a gradual rise in the content of SiO_2. The ettringite and C-S-H peak intensities of the CLSM samples with more BA content are lower. This is because the amount of fly ash is negatively correlated with the amount of BA, and the decline in fly ash leads to a decline in the level of hydration reaction that produces C-S-H, resulting in a decrease in the intensity of the C-S-H peak.

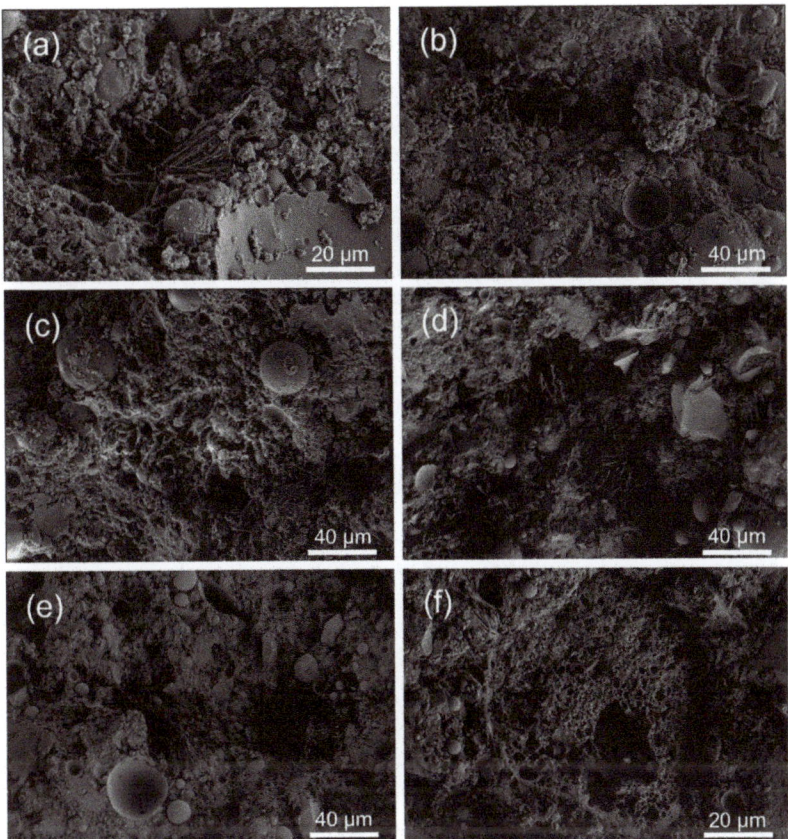

Figure 10. SEM images of prepared CLSM mixtures. (**a**,**b**) Blank, (**c**) BAI30, (**d**) BAI70, (**e**) BAII30, (**f**) BAII70.

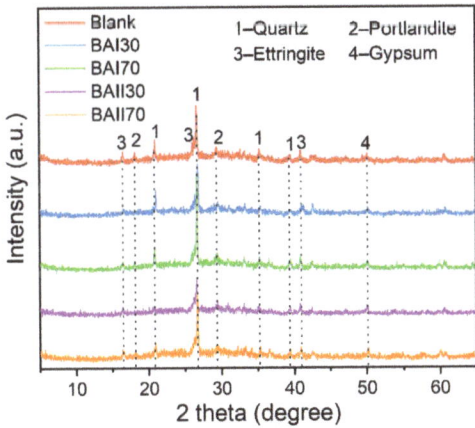

Figure 11. XRD patterns of prepared CLSM mixtures at 28 d: Blank, BAI30, BAI70, BAII30, and BAII70.

4. Conclusions

The present work shows that the use of five different by-products and cement to produce CLSM is very feasible. Furthermore, the effects of increasing the amount of BA and increasing the BA ball milling time on the fresh and hardened properties of CLSM are investigated, respectively. The performance of some CLSM was enhanced by increasing BA content and increasing the BA ball milling time. The conclusions of this study can be summarized as follows.

(1) The flowability, bleeding, compressive strength, setting time, density, porosity, and absorption of CLSM met the specification and requirements of ACI Committee 229. The flowability and bleeding of CLSM decreased with the increase in BA content and ball milling time. However, the flowability was still in the high flowability range, and the reduction in bleeding was favorable. Bleeding was reduced by 48% and 64% for BAI70 and BAII70, respectively. The density, porosity, and absorption of CLSM did not change significantly with the addition of BA and the change in ball-milling time. With the increase in BA content, the initial setting time and final setting time of CLSM declined significantly, and the final setting time of BAI70 and BAII70 decreased by 37% and 36%, respectively.

(2) The addition of BA and the increase in ball-milling time improved the 3 d strength of CLSM. Compared with Blank, BAI70 and BAII70 increased by 48% and 93%, respectively, which was favorable for structural fills. With the increase in the mass ratio, the 7 d and 28 d strength showed a trend of first declining and then increasing, but the fluctuation was not remarkable. The RE values of all CLSM mixtures were all greater than 1, which is suitable for structural fills that require a higher strength.

(3) It was observed in the SEM images that the BA-containing CLSM exhibited an open microstructure with high porosity. As revealed by XRD, with the rise in BA content, the quartz peaks of CLSM samples were enhanced, whereas the intensities of ettringite and C-S-H peaks were reduced.

(4) The production of CLSM with coal-based solid wastes as raw materials is feasible in terms of engineering performance, cost, and environmental impact. In the future, research on cement-free CLSM for total solid wastes should be strengthened to further reduce costs. Additionally, considering the location, utilization rate, and economy of raw materials, the feasibility of large-scale production and application of the CLSM prepared in this work for underground filling should be discussed.

Author Contributions: Conceptualization, T.C., N.Y. and S.W.; methodology, N.Y. and S.W.; validation, T.C. and S.W.; formal analysis, T.C., X.Z. and S.W.; investigation, T.C. and S.W.; resources, N.Y., X.H. and D.W.; data curation, T.C., N.Y. and S.W.; writing—original draft preparation, T.C., N.Y. and S.W.; writing—review and editing, T.C., N.Y., X.Z., X.Y. and S.W.; visualization, T.C., N.Y. and S.W.; supervision, N.Y.; project administration, N.Y. and D.W.; funding acquisition, N.Y.; All authors have read and agreed to the published version of the manuscript.

Funding: This work was supported by the National Key Research and Development Program of China (2019YFC1904304), the University-Industry Collaborative Education Program of the Ministry of Education of China (202101255047), and the Fundamental Research Funds for the Central Universities (2022YQHH09).

Institutional Review Board Statement: Not applicable.

Informed Consent Statement: Not applicable.

Data Availability Statement: The data presented in this study are available on request from the corresponding author.

Acknowledgments: The authors are grateful to the School of Chemical and Environmental Engineering, China University of Mining and Technology (Beijing) for the experimental equipment and technical support.

Conflicts of Interest: The authors declare no conflict of interest.

References

1. *ACI 229R-13*; Report on Controlled Low-Strength Materials. American Concrete Institute: Farmington Hills, MI, USA, 2013.
2. Somboonyanon, P.; Halmen, C. Seismic Behavior of Steel Pipeline Embedded in Controlled Low-Strength Material Subject to Reverse Slip Fault. *J. Pipeline Syst. Eng.* **2021**, *12*, 04021025. [CrossRef]
3. Alizadeh, V. Analytical study for allowable bearing pressures of CLSM bridge abutments. *Transp. Geotech.* **2019**, *21*, 100271. [CrossRef]
4. Do, T.M.; Kim, H.; Kim, M.; Kim, Y. Utilization of controlled low strength material (CLSM) as a novel grout for geothermal systems: Laboratory and field experiments. *J. Build. Eng.* **2020**, *29*, 101110. [CrossRef]
5. Bassani, M.; Bertola, F.; Bianchi, M.; Canonico, F.; Marian, M. Environmental assessment and geomechanical properties of controlled low-strength materials with recycled and alternative components for cements and aggregates. *Cem. Concr. Compos.* **2017**, *80*, 143–156. [CrossRef]
6. Mneina, A.; Soliman, A.M.; Ahmed, A.; El Naggar, M.H. Engineering properties of Controlled Low-Strength Materials containing Treated Oil Sand Waste. *Constr. Build. Mater.* **2018**, *159*, 277–285. [CrossRef]
7. Lee, N.K.; Kim, H.K.; Park, I.S.; Lee, H.K. Alkali-activated, cementless, controlled low-strength materials (CLSM) utilizing industrial by-products. *Constr. Build. Mater.* **2013**, *49*, 738–746. [CrossRef]
8. Hung, C.C.; Wang, C.C.; Wang, H.Y. Establishment of the Controlled Low-Strength Desulfurization Slag Prediction Model for Compressive Strength and Surface Resistivity. *Appl. Sci.* **2020**, *10*, 5674. [CrossRef]
9. Do, T.M.; Kang, G.O.; GO, G.H.; Kim, Y.S. Evaluation of Coal Ash-Based CLSM Made with Cementless Binder as a Thermal Grout for Borehole Heat Exchangers. *J. Mater. Civ. Eng.* **2019**, *31*, 04019072. [CrossRef]
10. Katz, A.; Kovler, K. Utilization of industrial by-products for the production of controlled low strength materials (CLSM). *Waste Manag.* **2004**, *24*, 501–512. [CrossRef]
11. Xiao, R.; Polaczyk, P.; Jiang, X.; Zhang, M.M.; Wang, Y.H.; Huang, B.S. Cementless controlled low-strength material (CLSM) based on waste glass powder and hydrated lime: Synthesis, characterization and thermodynamic simulation. *Constr. Build. Mater.* **2021**, *275*, 122157. [CrossRef]
12. Dueramae, S.; Sanboonsiri, S.; Suntadyon, T.; Aoudta, B.; Tangchirapat, W.; Jongpradist, P.; Pulngern, T.; Jitsangiam, P.; Jaturapitakkul, C. Properties of lightweight alkali activated controlled Low-Strength material using calcium carbide residue–Fly ash mixture and containing EPS beads. *Constr. Build. Mater.* **2021**, *297*, 123769. [CrossRef]
13. Zhen, G.Y.; Lu, X.Y.; Zhao, Y.C.; Niu, J.; Chai, X.L.; Su, L.H.; Li, Y.Y.; Liu, Y.; Du, J.R.; Hojo, T.; et al. Characterization of controlled low-strength material obtained from dewatered sludge and refuse incineration bottom ash: Mechanical and microstructural perspectives. *J. Environ. Manag.* **2013**, *129*, 183–189. [CrossRef]
14. Cheng, L.C.; Qin, Y.L.; Li, X.W.; Zhao, X.Y. A Laboratory and Numerical Simulation Study on Compression Characteristics of Coal Gangue Particles with Optimal Size Distribution Based on Shape Statistics. *Math. Probl. Eng.* **2020**, *2020*, 8046156. [CrossRef]
15. Jala, S.; GoyalL, D. Fly ash as a soil ameliorant for improving crop production—A review. *Bioresour. Technol.* **2006**, *97*, 1136–1147. [CrossRef]
16. Dmitrienko, M.A.; Strizhak, P.A. Environmentally and economically efficient utilization of coal processing waste. *Sci. Total Environ.* **2017**, *598*, 21–27. [CrossRef]
17. Siddique, R. Utilization of coal combustion by-products in sustainable construction materials. *Resour. Conserv. Recycl.* **2010**, *54*, 1060–1066. [CrossRef]
18. Li, J.Y.; Wang, J.M. Comprehensive utilization and environmental risks of coal gangue: A review. *J. Clean. Prod.* **2019**, *239*, 117946. [CrossRef]
19. Miao, C.; Liang, L.X.; Zhang, F.; Chen, S.M.; Shang, K.X.; Jiang, J.L.; Zhang, Y.; Ouyang, J. Review of the fabrication and application of porous materials from silicon-rich industrial solid waste. *Int. J. Miner. Metall. Mater.* **2022**, *29*, 424–438. [CrossRef]
20. Gong, Y.B.; Sun, J.M.; Zhang, Y.M.; Zhang, Y.F.; Zhang, T.A. Dependence on the distribution of valuable elements and chemical characterizations based on different particle sizes of high alumina fly ash. *Fuel* **2021**, *291*, 120225. [CrossRef]
21. Zhang, Y.X.; Zheng, X.Z.; Cai, W.J.; Liu, Y.; Luo, H.L.; Guo, K.D.; Bu, C.J.; Li, J.; Wang, C. Key drivers of the rebound trend of China's CO_2 emissions. *Environ. Res. Lett.* **2020**, *15*, 104049. [CrossRef]
22. Volokitin, G.G.; Skripnikova, N.K.; Volokitin, O.G.; Lutsenko, A.V.; Shekhovtsov, V.V.; Litvinova, V.A.; Semenovykh, M.A. Bottom Ash Waste Used in Different Construction Materials. *IOP Conf. Ser. Mater. Sci. Eng.* **2017**, *189*, 012013. [CrossRef]
23. Wang, Y.L.; Zhao, Y.Q.; Han, Y.S.; Zhou, M. The Effect of Circulating Fluidised Bed Bottom Ash Content on the Mechanical Properties and Drying Shrinkage of Cement-Stabilised Soil. *Materials* **2022**, *15*, 14. [CrossRef]
24. Liu, R.; Vail, M.; Koohbor, B.; Zhu, C.; Tang, C.S.; Xu, H.; Shi, X.C. Desiccation cracking in clay-bottom ash mixtures: Insights from crack image analysis and digital image correlation. *Bull. Eng. Geol. Environ.* **2022**, *81*, 139. [CrossRef]
25. Pormmoon, P.; Abdulmatin, A.; Charoenwaiyachet, C.; Tangchirapat, W.; Jaturapitakkul, C. Effect of cut-size particles on the pozzolanic property of bottom ash. *J. Mater. Res. Technol.* **2021**, *10*, 240–249. [CrossRef]
26. González-Fonteboa, B.; Carro-López, D.; Brito, J.D.; Martínez-Abella, F.; Seara-Paz, S.; Gutiérrez-Mainar, S. Comparison of ground bottom ash and limestone as additions in blended cements. *Mater. Struct.* **2017**, *50*, 84. [CrossRef]
27. Mazurkiewicz, M.; Tkaczewska, E.; Pomykala, R.; Uliasz-Bochenczyk, A. Preliminary determination of the suitability of slags resulting from coal gasification as a pozzolanic raw material. *Gospod. Surowcami Min.* **2012**, *28*, 5–14. [CrossRef]

28. Kaliyavaradhan, S.K.; Ling, T.C.; Guo, M.Z. Upcycling of wastes for sustainable controlled low-strength material: A review on strength and excavatability. *Environ. Sci. Pollut. Res.* **2022**, *29*, 16799–16816. [CrossRef]
29. Ling, T.C.; Kaliyavaradhan, S.K.; Poon, C.S. Global perspective on application of controlled low-strength material (CLSM) for trench backfilling—An overview. *Constr. Build. Mater.* **2018**, *158*, 535–548. [CrossRef]
30. Zhang, Q.L.; Wu, X.M. Performance of cemented coal gangue backfill. *J. Cent. South Univ.* **2007**, *14*, 216–219. [CrossRef]
31. Yang, X.B.; Yan, Z.P.; Yin, S.H.; Gao, Q.; Li, W.G. The Ratio Optimization and Strength Mechanism of Composite Cementitious Material with Low-Quality Fly Ash. *Gels* **2022**, *8*, 151. [CrossRef]
32. ASTM D6103-17; Standard Test Method for Flow Consistency of Controlled Low Strength Material (CLSM). ASTM International: West Conshohocken, PA, USA, 2017. Available online: https://www.astm.org/d6103_d6103m-17.html (accessed on 18 June 2022).
33. GB/T 50080-2016; Standard for Test Method of Performance on Ordinary Fresh Concrete. China National Standard, Ministry of Construction: Beijing, China, 2016.
34. GB/T 50081-2019; Standard for Test Method of Concrete Physical and Mechanical Properties. China National Standard, Ministry of Construction: Beijing, China, 2019.
35. GB/T 1346-2011; Test Methods for Water Requirement of Normal Consistency, Setting Time and Soundness of the Portland Cement. China National Standard, General Administration of Quality Supervision, Inspection and Quarantine: Beijing, China, 2011.
36. ASTM D6023-16; Standard Test Method for Density (Unit Weight), Yield, Cement Content, and Air Content (Gravimetric) of Controlled Low-Strength Material (CLSM). ASTM International: West Conshohocken, PA, USA, 2016. Available online: https://www.astm.org/d6023-16.html (accessed on 18 June 2022).
37. Yang, T.; Zhu, H.J.; Zhang, Z.H.; Gao, X.; Zhang, C.S.; Wu, Q.S. Effect of fly ash microsphere on the rheology and microstructure of alkali-activated fly ash/slag pastes. *Cem. Concr. Res.* **2018**, *109*, 198–207. [CrossRef]
38. Hwang, C.L.; Chiang, C.H.; Huynh, T.P.; Vo, D.H.; Jhang, B.J.; Ngo, S.H. Properties of alkali-activated controlled low-strength material produced with waste water treatment sludge, fly ash, and slag. *Constr. Build. Mater.* **2017**, *135*, 459–471. [CrossRef]
39. Chen, C.G.; Sun, C.J.; Gau, S.H.; Wu, C.W.; Chen, Y.L. The effects of the mechanical–chemical stabilization process for municipal solid waste incinerator fly ash on the chemical reactions in cement paste. *Waste Manag.* **2013**, *33*, 858–865. [CrossRef] [PubMed]
40. Li, H.; Chen, Y.; Cao, Y.; Liu, G.J.; Li, B.Q. Comparative study on the characteristics of ball-milled coal fly ash. *J. Therm. Anal. Calorim.* **2016**, *124*, 839–846. [CrossRef]
41. Rani, R.; Jain, M.K. Effect of bottom ash at different ratios on hydraulic transportation of fly ash during mine fill. *Powder Technol.* **2017**, *315*, 309–317. [CrossRef]
42. Yu, J.; Li, G.Y.; Leung, C.K.Y. Hydration and physical characteristics of ultrahigh-volume fly ash-cement systems with low water/binder ratio. *Constr. Build. Mater.* **2018**, *161*, 509–518. [CrossRef]
43. Andrade, L.B.; Rocha, J.C.; Cheriaf, M. Influence of coal bottom ash as fine aggregate on fresh properties of concrete. *Constr. Build. Mater.* **2009**, *23*, 609–614. [CrossRef]
44. Kim, M.H.; Choi, S.J. An Experimental Study on the Properties of Concrete using High Volume of Coal Ash. *Archit. Res.* **2002**, *4*, 39–44.
45. Singh, M.; Siddique, R. Strength properties and micro-structural properties of concrete containing coal bottom ash as partial replacement of fine aggregate. *Constr. Build. Mater.* **2014**, *50*, 246–256. [CrossRef]
46. Ghanad, D.A.; Soliman, A.; Godbout, S.; Palacios, J. Properties of bio-based controlled low strength materials. *Constr. Build. Mater.* **2020**, *262*, 120742. [CrossRef]
47. Singh, M.; Siddique, R. Properties of concrete containing high volumes of coal bottom ash as fine aggregate. *J. Clean. Prod.* **2015**, *91*, 269–278. [CrossRef]
48. Le, N.H.; Razakamanantsoa, A.; Nguyen, M.; Phan, V.T.; Dao, P.; Nguyen, D.H. Evaluation of physicochemical and hydromechanical properties of MSWI bottom ash for road construction. *Waste Manag.* **2018**, *80*, 168–174. [CrossRef]
49. Li, E.P.; Chen, H.Y.; Huang, F.L.; Tian, S.Q.; Yu, Z.Y. The bottom ash from municipal solid waste and sewage sludge co-pyrolysis technology: Characteristics and performance in the cement mortar and concrete. *IOP Conf. Ser. Earth Environ. Sci.* **2020**, *585*, 012091. [CrossRef]
50. Kwon, W.T.; Kim, B.I.; Kim, Y.; Kim, S.R.; Ha, S.W. Characterization of Power Plant Bottom Ash and its Application to Cement Mortar. *Mater. Sci. Forum* **2009**, *620–622*, 221–224. [CrossRef]
51. Ghanad, D.A.; Soliman, A.M. Bio-based alkali-activated controlled low strength material: Engineering properties. *Constr. Build. Mater.* **2021**, *279*, 122445. [CrossRef]
52. Kim, Y.S.; Do, T.M.; Kim, H.K.; Kang, G. Utilization of excavated soil in coal ash-based controlled low strength material (CLSM). *Constr. Build. Mater.* **2016**, *124*, 598–605. [CrossRef]
53. Jang, J.G.; Lee, N.K.; Lee, H.K. Fresh and hardened properties of alkali-activated fly ash/slag pastes with superplasticizers. *Constr. Build. Mater.* **2014**, *50*, 169–176. [CrossRef]
54. Lee, N.K.; Lee, H.K. Setting and mechanical properties of alkali-activated fly ash/slag concrete manufactured at room temperature. *Constr. Build. Mater.* **2013**, *47*, 1201–1209. [CrossRef]
55. Park, S.M.; Lee, N.K.; Lee, H.K. Circulating fluidized bed combustion ash as controlled low-strength material (CLSM) by alkaline activation. *Constr. Build. Mater.* **2017**, *156*, 728–738. [CrossRef]

Article

Study on the Hydration and Microstructure of B and B/Na Ion-Doped Natural Hydraulic Lime Composed with Silica Fume/Fly Ash

Yanbo Zhang, Ze Liu *, Jixiang Wang, Conghao Shao, Jiaxing Li and Dongmin Wang

School of Chemical and Environmental Engineering, China University of Mining & Technology, Beijing 100083, China
* Correspondence: lzk1227@sina.com

Abstract: Natural hydraulic lime (NHL) has drawn much attention due to its environmentally friendly nature. The characteristics of both hydraulic and pneumatic components make it a potential substitute for Portland cement in surface decoration and ancient building restoration. In this study, both doping and mixing with supplementary cementitious materials were investigated. Two types of NHL3.5 were fabricated through calcination at 1200 °C with B and B/Na doping, respectively. It is noted that B ion doping is beneficial to the early compressive strength of the specimens, and B/Na doping is beneficial to the later compressive strength of the specimens. The observed outcome is that the compressive strengths of B and coupled B/Na doped NHL3.5 are higher than the blank sample due to the appearance of α'-C2S. Thereafter, the blank and doping NHL were incorporated with fly ash and silica fume. The incorporation of fly ash and silica fume could enhance the early and late hydration rate. Of the two, silica fume shows more pozzolanic effect in the early age. In the supplementary cementitious materials dosed group, pozzolanic dominates the hydration process.

Keywords: natural hydraulic lime; ion doping; supplementary cementitious materials; hydration mechanism

1. Introduction

Hydraulic lime (HL) [1], one of the most historic cementitious materials, has recently drawn wide attention in both the restoration of ancient buildings and relics and the field of architecture. Traced back to the ancient Roman and Greek periods, people used lime-based cementitious materials (LCMs) as the main building binder [2,3]. It was not until the appearance of Portland cement and its promotion in the 18th century that lime-based cementitious materials were phased out of history [4]. In recent years, with the development of the concept of "green building materials", LCMs have gradually regained the focus of researchers worldwide. Compared with the calcination temperature of 1250~1450 °C for cementitious materials, the calcination temperature of HL is lower, usually 800~1200 °C, which greatly reduces the carbon emission in the production process [5,6]. HL also absorbs CO_2 from the air during the hardening process, which further reduces energy consumption and carbon emissions. At the same time, the choice of raw materials for the preparation of hydraulic lime is wider: marl [7], chalk [8], ginger nut [6], even dolomite [9] and other low-grade silica-calcium minerals that cannot meet the calcination conditions of the cement industry can be used as the main raw materials for natural hydraulic lime (NHL), which enhances the comprehensive utilization of minerals. Additionally, due to the wide range of raw materials, NHL's production conditions and application methods can be adapted to local conditions and reduce transportation costs to a certain extent. In conclusion, NHL has great potential for application in the construction field of ancient building restoration [10] and surface decorative mortars [11], coatings, etc. However, there are some problems in the application of hydraulic lime, such as cracking and salt petering [12]. In the field of ancient building restoration, the long early setting time and low mechanical strength of hydraulic lime also seriously restrict its practical application [13]. The key to solving these problems is to improve the hydration activity of NHL materials and to conduct more in-depth research on the hydration mechanism and the working performance mechanism.

The main components of NHL include the gaseous hard component calcium hydroxide ($Ca(OH)_2$) and the hydraulic component ($2CaO\text{-}SiO_2$, C_2S). As reported, there are five types of polymorphs in C_2S, named α, $\alpha'H$, $\alpha'L$, β, and γ, respectively [14]. Where α, $\alpha'H$, and $\alpha'L$ phase are considered with high hydraulic reactivity, the instability at room temperature strongly restricts its application [15]. With respect to β phase, it is the main crystalline phase of C_2S in cement clinker and commercial NHL [16]. At the same time, γ phase is commonly considered the most stable polymorph and hardly shows hydraulic reactivity [17]. Doping with impurity ions is one of the methods commonly used to improve the hydraulic activity of C_2S. Na^+, Fe^{3+}, K^+, Mg^{2+}, and other minor ions can be doped onto the C_2S crystal to replace Ca^{2+} or $[SiO_4]^{4-}$, which hinders the transformation of C_2S into a rhombohedral crystal system during the cooling process [18,19].

It is well known that supplementary cementitious materials (SCMs) have a wide range of applications in the cement and hydraulic lime industry [20]. According to BSEN459:1-2010, NHLs can be referred as HLs when SCMs are added. SCMs have the following advantages: (i) reduced clinker usage; (ii) improved workability of the binder; (iii) pozzolanic activity can improve the mechanical strength of the matrix, etc. [21,22]. The commonly used SCMs are blast furnace slag (BFS) [23], fly ash (FA) [24], silica fume (SF) [25], and metakaolin (MK) [26]. In recent decades, the development of composite cement properties and microstructures has been a high research priority.

In this study, the NHL containing highly active α and α'-C_2S was prepared by first mixing limestone and diatomite according to the calcium-silica component requirements of BSEN459:1-2010 [27] for NHL3.5, then doping with a small amount of stabilizer (B, B/Na ions), followed by calcination and digestion. Then the fabricated NHL was complexed with FA and SF in different proportions and prepared as pastes with a fixed w/c ratio of 0.55. The hydration mechanism, microstructural changes, and mechanical properties of HL-based cementitious materials under the composite effect of dynamic material of SCMs and ion doping were investigated. In this paper, ion doping and SCMs were used together, and the hydration reaction difference of C2S with different crystal forms was explored. The early hydration reaction speed of NHL was improved and the application of NHL was expanded.

2. Materials and Methods

2.1. Materials

The NHL used in this study was fabricated in the laboratory. The raw materials used to prepare it are limestone from Liaoning Province and diatomite from Jilin Province, China. SCMs are fly ash and silica fume from Shanxi Province, China.

The chemical composition of raw materials is shown in Table 1. The raw materials were heated to 1000 °C to determine the loss on ignition (L.O.I) of raw materials, and the L.O.I of limestone at 1000 °C was 43.09 wt%, corresponding to a $CaCO_3$ content of about 97.9%. Meanwhile, the content of MgO is about 3 wt%, which, according to XRD results (Figure 1), mainly corresponds to dolomite ($CaMg(CO_3)_2$) in the raw material. According to the calculation, the Cementation Index (CI) value of the final product was 0.85 when adding 12% diatomite, which theoretically allows the production of natural hydraulic lime with high hydraulic properties.

$$CI = \frac{((2.8 \times \%SiO_2) + (1.1 \times \%Al_2O_3) + (0.7 \times \%Fe_2O_3))}{(\%CaO + 1.4 \times \%MgO)} \quad (1)$$

Table 1. Chemical components of the raw materials (wt%).

Materials	CaO	SiO$_2$	Al$_2$O$_3$	Fe$_2$O$_3$	MgO	SO$_3$	Others	L.O.I
limestone	48.33	3.61	0.81	0.73	3.00	0.05	0.38	43.09
Diatomite	0.32	86.20	2.03	1.93	0.26	0.29	0.82	8.14
Fly ash	2.88	50.54	26.86	5.95	0.74	0.52	5.93	6.59
Silica fume	0.18	92.70	0.14	0.04	0.64	0.59	0.84	4.87

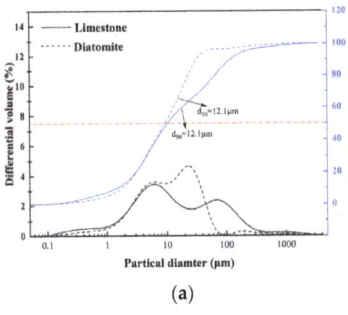
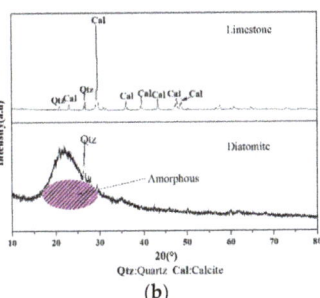

(a) (b)

Figure 1. (a) Particle size diameter distribution of raw material, (b) XRD patterns of the raw materials.

In this experiment, fly ash (FA) and silica fume (SF) were used as SCMs, where FA is mainly composed of CaO, Al$_2$O$_3$, Fe$_2$O$_3$, and other oxides. The content of CaO in FA is about 2.88%, indicating that it is low-calcium silicon FA. SF is acknowledged as highly reactive pozzolanic material, and its activity is mainly derived from the amorphous SiO$_2$, which accounts for 92.7% of the total content.

2.2. Preparation

According to the component requirements of BSEN459-1:2010 for NHL3.5, the Ca(OH)$_2$ content should not be less than 25%. Thus, the designed NHL3.5 consists of 75% C$_2$S and 25% Ca(OH)$_2$. First, 12 wt% diatomite was mixed with 88 wt% limestone as the calcined raw materials, added to 1.5 wt% B$_2$O$_3$, and 1.5 wt% B$_2$O$_3$ and Na$_2$CO$_3$ (in qual mass ratio) were mixed as stabilizers, respectively, and the mixture was ball-milled for 30 min until the particle size reached below 0.4 mm. As can be seen from Figure 2, the mixed powder was put into the muffle furnace, calcined at 1200 °C, with a heating rate of 5 °C/min and a holding time of 120 min. Then, 30% water by weight of the calcined product was weighed and mixed with the calcined product until the f-CaO was completely converted to Ca(OH)$_2$, which could be observed through the exothermic condition, and was dried at 80 °C after digestion for approximately 20 min. The final product was obtained after 30 min of grinding. The granularity of the final product reached 0.2 mm, the residual value of the sieve did not exceed 2% of the total mass of the sieved material, and the residual value of the sieve of 0.09 mm did not exceed 15% of the total mass of the sieved material.

Referring to the cement net test methods GBT1346-2011, 500 g of prepared NHL were added into the net paste mixer, water was added at a water-binder ratio of 0.55, the mixer was started and mixed at low speed for 120 s and then stopped for 15 s, and the slurry on the blade and pot wall was scraped into the pot and mixed at high speed for 120 s. It was poured into a 30 × 30 × 30 mm mold for molding. The molding temperature was 20 °C and the humidity was 60%. After 48 h, the demolded specimens were placed in the same environment and continued to cure to specific ages. At specific ages, the mechanical properties of the specimens were determined. Meanwhile, the hydration process was terminated by grounding into powder and soaking in ethanol. Before microscopic characterization, the samples were dried in a vacuum oven at 65 °C for 24 h. The mixture ratio is shown in Table 2.

Figure 2. Schematic diagram of preparation process of natural hydraulic lime.

Table 2. Mixture mass (g) ratio of the materials in the experiment.

	Sample	NHL	B1.5	BNa1.5	FA	SF
Blank	NHL	500				
	F1	450			50	
	F2	400			100	
	S1	450				50
	S2	400				100
B-ion doping	B1.5		500			
	F1B1.5		450		50	
	F2B1.5		400		100	
	S1B1.5		450			50
	S2B1.5		400			100
B/Na doping	BNa1.5			500		
	F1BNa1.5			450	50	
	F2BNa1.5			400	100	
	S1BNa1.5			450		50
	S2BNa1.5			400		100

2.3. Test Methods

The components of raw materials were measured by PANalytical Axios XRF (X-ray Fluorescence Spectrometer, PANalytical, Almelo, The Netherland). Smatlab X-ray diffraction (XRD, Rigaku Corporation, Tokyo, Japan) was used to determine the mineral composition of the raw material and specimens, with 40 kV and 40 mA, CuKa1 radiation. Furthermore, 10°/min and 1°/min increment and 1 s·step-1 sweep from 10° to 80° 2 h were adapted, respectively. Meanwhile, the Rietveld method was used for XRD data analysis to quantify the mineral phase content.

Tam Air 08 isothermal calorimeter (TA Instrument, New Castle, DE, USA), was used to determine the hydration heat of specimens. At 20 °C, the specimens were weighed and placed in 20 mL of ampoule according to the proportion, and water was added to them according to the W/B ratio of 0.55. They were stirred quickly and put into the calorimeter.

Fourier transform infrared spectroscopy (FT-IR) spectra were carried out in the range of 400–4000 cm^{-1} to determine chemical bonding and crystal structure changes. Microscopic morphology and structure were determined by scanning electron microscope (SEM, Hitachi Regulus8100, Tokyo, Japan). The pore structure of the matrix is characterized by AutoPore IV 9500 Mercury intrusion porosimetry (MIP, Micromeritics Instrument, Norcross, GA, USA) in 30,000 psi. The mechanical properties of the specimens were measured on the same test machine (SANS CMT5105, Shenzhen, China) at a loading rate of 2400 ± 200 N/s.

3. Results
3.1. Mineral Content and Crystal Structure

For quantitative analysis, 20% rutile was mixed into the specimens as an internal calibrator. The XRD identified the five main mineral phases: portlandite, α'-C_2S, β-C_2S, rutile (used as an internal standard), calcite, and magnesite, as shown in Table 3. The GSAS software was used to make a refined fit to the test data. The final fitting results are shown in Figure 3. The fit difference Rwp of each specimen is less than 10%, which indicates a good matching result. The main error is from α'-C_2S and β-C_2S, and the characteristic peaks of the two phases overlapped extensively.

According to the results of quantitative analysis, the content of magnesite is about 3.27~4.31%. It is in accordance with BS EN459-1 that the free MgO content of NHL should be less than 7%, and, after mixing with SCMs, the content of magnesite in the system will be further reduced. At the same time, Mg ions will participate in constituting other crystalline phases and amorphous phases in the system. The content of magnesite in all three types of NHL prepared in this experiment did not exceed the regulation, and magnesite did not affect the system significantly in the short term.

From Figure 3, it can be seen that in the undoped sample, the diffraction peaks of C_2S are mainly β phase, and in the B-doped sample there is an obvious crystalline transformation, and the diffraction peaks of α'-C_2S appear obviously at about $2\theta = 33°$. In addition, the diffraction peak at $2\theta = 33°$ tends to shift to a higher angle [28], which may be attributed to the replacement of small radius B ions with large radius Si ions in the doping process, and thus the size of α'-C_2S is reduced [29]. As can be seen from the composition of the specimens in Figure 3d, the content of C_2S increases slightly after the incorporation of the stabilizer.

When B ions were doped, the C_2S crystalline phase in the prepared NHL was α'-phase with a content of 42.4 wt%, while 6.94 wt% of β-phase C_2S was also present in the system, and when B/Na ions were doped, the content of α'-C_2S decreased to 38.56 wt%, whereas the content of the amorphous phase increased. This phenomenon is in agreement with the results of Chen, L. et al. [30]. Similarly, Álvarez-Pinazo et al. attributed this mainly to the presence of excess dopant in the belite as well as cell defects [31]. Additionally, the Na ions are said to have an antagonistic effect with B ions and stabilize more C_2S into β phase.

Table 3. ICCD-PDF and ICSD collection codes for all phases used for Rietveld refinements.

Phase	Space Group	ICSD Code	PDF Code
α'-C_2S	Pnma	81097	49-1674
β-C_2S	P121/N1	963	33-0302
Portlandite	P-3M1	15471	44-1481
Rutile	P42/mnm	9161	21-1276
Calcite	P3221	174	78-2315
Periclase	Fm-3m	9863	45-0946

In addition to their stabilizing effect on belite, B and Na ions are often used as mineralizing agents in the cement industry. Their addition can enhance the burnability of cement, lower the phase formation temperature of C_2S, C_3S, etc. [32,33], and improve the calcination effect. Therefore, after doping with minor ions, the denseness of calcination products rises and the crystal structure of CaO becomes more stable, which may affect its reaction with water and thus adversely affect its long-term stability [34]. It is inclined to be not easily pulverized during the digestion process, which is objectively manifested by the rise in particle size of NHL, as shown in Figure 4. So the influence of doping ions can be concluded as doping B ion is more beneficial to the formation of α'-C_2S, while doping B/Na ions lowered the content of α'-C_2S. Mean, while B/Na ion doping shows more mineralization effect which led to the decrease in the content of portlandite and an increase in the particle size of the sample.

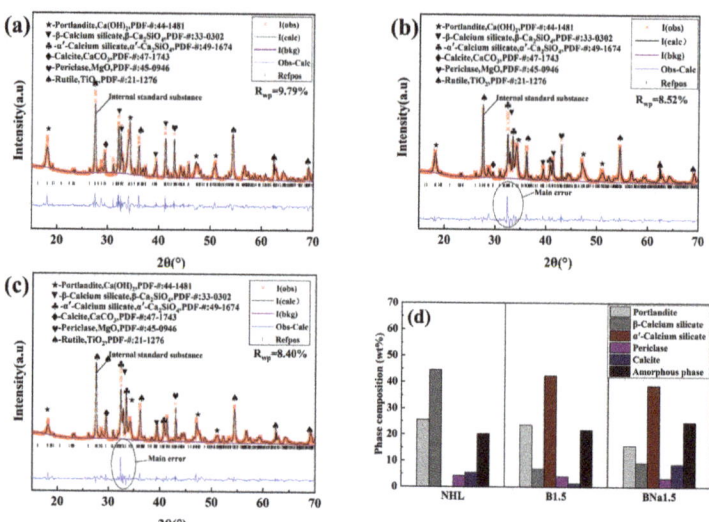

Figure 3. XRD fitting pattern of (**a**) NHL, (**b**) B1.5, and (**c**) BNa1.5; (**d**) mineral phase composition of the sample by Rietveld method.

It can be seen from Figure 5 that the characteristic bands near 842 cm^{-1}, 900 cm^{-1}, 1000 cm^{-1} in the undoped samples correspond to the stretching vibrations of the Si-O bond and can be identified as β-C_2S [35]. Meanwhile, new bands near 746 cm^{-1} and 1245 cm^{-1} can be observed in the doped samples, corresponding to the bending and stretching vibrations of [BO_3]$^{3-}$. According to the research, the structure of α'-C_2S stabilized by B ion can be defined as $Ca_{2-x}B_x(SiO_4)_{1-x}(BO_4)_x$, where the x is determined by the dosage of B ions as well as the ion types and quantity of raw material [36].

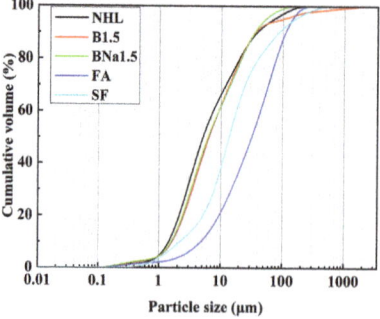

Figure 4. NHL, FA, and SF particle size distribution.

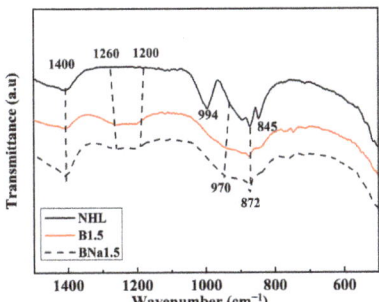

Figure 5. FT-IR spectra of laboratory-prepared NHL.

Na$^+$ ion is considered an efficient dopant to stabilize β-C$_2$S. It is generally agreed that Na$^+$ ion can replace Ca^{2+} in equivalent molar amounts in C$_2$S crystal [37]. For the samples with complex doping of B and Na ions, the characteristic spectra are similar to those of the single doped B ion specimens. Both [BO$_3$]$^{3-}$ as well as [BO$_4$]$^{5-}$ were present in the specimens. In the calcination process, due to the existence of different types of mineral phases, such as quartz phase, calcium oxide, calcite, etc., Na and B ions may occur in different degrees of solid solution with each mineral phase, which leads to the unequal doping of B and Na for C$_2$S [38].

The results of Rietveld refined cell parameters are shown in Table 4. The variation of the cell parameters of β-C$_2$S and α'-C$_2$S also shows that the cell volumes of β-C$_2$S and α'-C$_2$S appear to be reduced to different degrees in the single-doped specimens and the complex-doped specimens. The cell volume of β-C$_2$S is smaller in the single-doped specimens than in the complex-doped samples, while the cell volume of α'-C$_2$S is smaller in the complex-doped samples. This is mainly due to the replacement of the large radius Si and Ca ions with the smaller radius B and Na ions. The different doping rates of B and Na atoms in β-C$_2$S and α'-C$_2$S cause different degrees of cell distortion. This result is also consistent with the FT-IR results.

Table 4. Lattice parameters of β-C2S and α'-C2S modified by the Rietveld method (Å).

Samples	β-C$_2$S				α'-C$_2$S			
	V	a	b	c	V	a	b	c
NHL	345.74	5.51	6.76	9.32				
B1.5	337.96	5.50	6.64	9.27	345.08	6.84	5.46	9.25
BNa1.5	345.68	5.51	6.75	9.31	344.91	6.85	5.45	9.24

3.2. Hydration Heat

The effect of doping and SCMs on the hydration characteristics of NHL was further investigated. Meanwhile, 10% and 20% FA and SF were added to the three types of NHL, respectively. Generally, the hydration process could be divided into four periods: (I) Initial reaction, (II) Inducing period, (III) Acceleration period, and (IV) Deceleration period. During the first period of mixed powder contact with water, the Ca(OH)$_2$ and C$_2$S dissolve and release a lot of heat [39]. Then the C$_2$S reacted with water and generated C-S-H gel layers which next act as a diffusion barrier coated with unreacted particles and restricted the hydration and pozzolanic reaction. Similar processes also occurred on the surface of SCMs particles [40]. Comparing the hydration exothermic curves of three specimens, NHL, B1.5, and BNa1.5, it is known that doping with B and Na ions significantly enhanced the hydration exothermic rate of the specimens. As can be seen from Figure 6, the exothermic rate curve of the blank group showed a continuous decay trend, and there was no obvious exothermic peak even till 160 h. Compared with the single doped B specimens, the exothermic peak of the coupled B and Na group appeared later, and its exothermic peak appeared at about 120 h, followed by a slow decay, and the continuous exotherm still appeared at 220 h of hydration. This phenomenon is consistent with the strength development mode, which further indicates that the single-blended B has higher early strength, while the coupled B and Na have higher late strength.

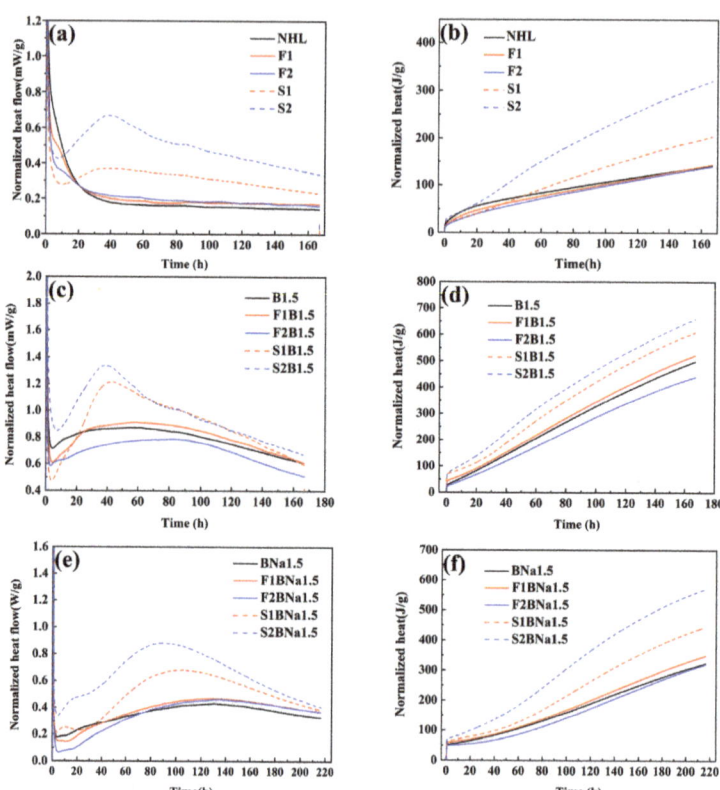

Figure 6. (a) Heat release rate curves of the blank group; (b) cumulative heat release of blank group; (c) heat release rate curves of B doping group; (d) cumulative heat release of B doping group; (e) the heat release rate curves of B/Na doping group; and (f) the cumulative heat release of B/Na doping group.

From Figure 6a, it can be seen that an obvious exothermic peak appeared at about 40 h after the incorporation of SF, and the peak intensity increased with the increase of SF mixing amount, which is due to the pozzolanic reaction of SF which generates a large amount of C-(A)-S-H gel and releases a large amount of heat. There is a slight decrease in the exothermic rate after the incorporation of fly ash, which indicates that FA is less involved in the early stage of hydration, while the relatively lower NHL content is another reason for the slower exotherm [41]. A similar phenomenon was also apparent in D. Zhang's study [42]. For the B-doped composite modified hydraulic lime, a significant increase in the hydration rate was observed for both specimens, relative to the blank group. As shown in Figure 6c, for the SF-doped specimens, an obvious exothermic peak appeared around 40 h, which can be identified as the exothermic peak generated by the pozzolanic reaction of SF after comparison with the blank group, which also indicates that the exotherm of pozzolanic reaction dominates the hydration in this stage. Thereafter, the exothermic curve gradually decreased, and at about 100 h, the curve of the SF-doped group was similar to the exothermic rate curve of the B1.5 specimen, indicating that the hydration of C_2S dominated at this stage and lasted until 160 h. In this group, the exothermic rate of the F1B1.5 specimen with 10% FA incorporation was higher than that of the B1.5 specimen (the exothermic rate curve became significantly steeper during the acceleration period), which may be attributed to: (i) The doping ions promoted the pozzolanic reaction of FA. (ii) The incorporation of FA increases the nucleation sites in the paste, which facilitates the nucleation growth of C-(A)-S-H gel. (iii) The "dilution effect" increases the contact between C_2S and water, so it can fully engage in the hydration reaction, which also leads to the enhancement of the heat of hydration [43]. When 20% FA was added, the clinker and calcium hydroxide content in the system is further reduced, which caused a further decline in the exothermic rate. The pattern of the coupled B and Na doping group is similar to that of B doping single. In this group, the exothermic peak of SF-doped specimens appeared later at about 80 h. In general, SF promotes the early exotherm of the specimens due to the high pozzolanic activity, while the relatively early low pozzolanic activity of FA declined the hydration rate of the specimens. This is in agreement with the early mechanical strength.

3.3. Mechanical Properties

The pozzolanic reaction consumes a large amount of $Ca(OH)_2$ in the system, and the resulting C-(A)-S-H gel is similar to the hydration products of C2S. The decrease in alkalinity reduces the hydration rate of C_2S, but the C-(A)-S-H gel induces the hydration reaction of C_2S in the long age, and the $Ca(OH)_2$ produced by the hydration reaction of C_2S also promotes the secondary pozzolanic reaction of SCMs, so the incorporation of SCMs delays the hydration reaction of C_2S in the early age, but promotes C_2S reaction from two aspects in the long age.

It is noticeable that the doping significantly increased the compressive strength of the specimens. At the early stage of curing, the compressive strength of the B-doped specimens was higher, reaching 18 MPa at 21 d, which was 5 MPa higher than that of the coupled B/Na-doped specimens and much higher than that of the undoped specimens, which was 1.6 MPa. After 21 d, the strength growth of the B-doped specimens slowed down, while the strength of the coupled specimens accelerated, reaching 22.1 MPa at 60 d. The above results indicated that the B-doped specimens were more favorable to the early strength of the specimens, while the coupled B/Na was more favorable to the later strength of the NHL.

For the blank group experiments, the incorporation of SF significantly enhanced the early strength of NHL, as shown in Figure 7. In the high alkalinity environment, amorphous SiO_2 with high activity within SF can react with $Ca(OH)_2$ more quickly to form C-(A)-S-H gel; thus, the early strength of the mortar developed better. The strength of the specimen with 20% silica fume reached 14.8 MPa at 28 d. However, at the late stage of curing, the strength of the specimen showed significant inversion shrinkage and severe cracking at 60 d, making its strength untestable. This could be attributed to the quick early hydration of the specimen. During the hydration process, the free water in the pore decreases continuously with the reaction and produces partial minor shrinkage, while the hydration products bond the unhydrated particles into a shelf structure, and refill the cracks and pore structure, forming a relatively stable structure to resist part of the shrinkage stress [44]. Meanwhile, the curing condition is also one of the reasons for cracking. Some researchers use water curing or steam curing to reduce the water loss rate of cementitious materials, thus reducing the loss of structural water. At the same time, in the environment with high humidity, a small amount of calcium hydroxide will also dissolve in water and repair the possible micro cracks. The C-S-H gel is more evenly dispersed. This curing method is often used for the cementitious materials with high hydration speed to avoid cracks [45].

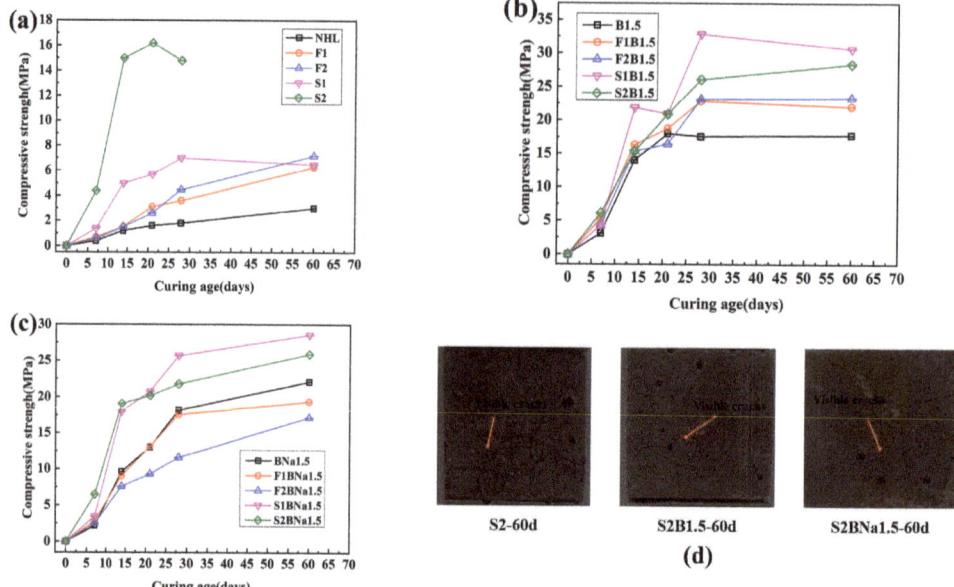

Figure 7. (a) Blank group, (b) B doping group, (c) B/Na doping group compressive strength, and (d) surface crack of the samples.

Due to the low early strength of NHL, it is not sufficient to resist the stresses due to shrinkage, thus producing cracks on the surface of the specimens, as shown in Figure 7d. The early strength of the FA-doped specimens in the blank group was lower compared to that of the specimens in the SF-doped group. The low early activity of FA is mainly related to its structure, which is a spherical particle covered with a glass layer that hinders the dissolution of the internal Si and Al phases. However, its strength is still slightly higher than that of the specimens without FA, which is mainly due to the interfiling of FA and lime particles, making the accumulation denser. In fact, at the initial stage, in addition to the degree of hydration, the filling effect and the surface properties of the particles also affect the strength of the specimens. The strength of the FA incorporated specimens continued to increase in the later stages of curing, as shown in Figure 7a. This is mainly due to the pozzolanic effect of FA of enhancing the strength of the specimens at the later stage [46]. In addition, the specimens in the FA dosed group were more structurally intact and did not show obvious cracks. On the one hand, the specimens in the early FA dosed group hydrated more slowly, thus the self-shrinkage was weaker and the shrinkage stress was low; on the other hand, the filling of FA between the particles optimized the structure of the specimens. In general, the SF-dosed specimens possessed higher early strength, but the mechanical properties deteriorated later due to self-shrinkage; the FA-dosed specimens had lower early strength but avoided microcracks in the specimens, and the pozzolanic effect of FA enhanced the mechanical properties of the specimens later. FA is considered to help reduce the cracking of cementitious materials.

After the doping of NHL, its composite modification with FA and SF resulted in some new effects on the specimens. First, for the specimens in the single B-doped group, the composite modification further improved the early mechanical properties of the specimens. The strengths of the specimens incorporated with 10% and 20% SF reached 32.9 MPa and 26.1 MPa at 28 d. The strengths of the specimens incorporated with 10% and 20% FA were 22.9 MPa and 23.2 MPa at 28 d, which were much higher than those of the blank group.

Two main types of hydration processes exist in this experiment. On the one hand, there is more α'-C_2S in the doped specimens, and its faster hydration rate at the early stage produces part of the C-(A)-S-H gel; on the other hand, the dissolution and polymerization process of Si and Al phases in FA generates another part of C-(A)-S-H. These two processes together determine the strength of the composite modified specimens, as Equations (2a–c) show.

$$\alpha'\text{-}2CaO \cdot SiO_2 + nH_2O \rightarrow mCaO \cdot SiO_2 \cdot kH_2O + (2-m)Ca(OH)_2 \qquad (2a)$$

$$xCa(OH)_2 + SiO_2 + nH_2O \rightarrow mCaO \cdot SiO_2 \cdot (x+n)H_2O \qquad (2b)$$

$$xCa(OH)_2 + Al_2O_3 + nH_2O \rightarrow mCaO \cdot Al_2O_3 \cdot (x+n)H_2O \qquad (2c)$$

These two hydration processes are not completely independent. In some previous studies, there are two main explanations for the interaction between the two processes. One explanation suggests that the promotion of alkalinity has a facilitated effect on the activity of fly ash [47]. The hydration process of the paste is accompanied by an increase in pH, and the increase in pH accelerates the erosion of the SCMs, thus more Si and Al phases are dissolved, which enhances the pozzolanic reaction. Another explanation is that the presence of SCMs has a facilitated effect on the hydration of C_2S [48]. A.M. Sharar et al. [49]. suggested that $Ca(OH)_2$ covers the surface of C_2S during hydration, which has a hindering effect on hydration. In contrast, the pozzolanic reaction consumes $Ca(OH)_2$, thus increasing the hydration activity of C_2S. It should be noted that both promotion mechanisms may exist simultaneously in the hydration process.

Finally, for the B-doped activation group, although some inversion of strength still occurred in the SF-doped specimens at the late stage of hydration, the inversion was improved relative to the blank group specimens, as can also be seen in Figure 7d, where the cracks on the surface of the S2B1.5 specimens were significantly reduced.

In general, SF with high volcanic ash activity can significantly promote the early hydration of NHL, while the effect of FA is not obvious at the early age and even reduces the early strength of NHL at 20% more incorporation. However, the effect of FA on the doped sample can significantly enhance the early performance of the matrix, owing to the interaction of the two hydration mechanisms. In addition, doping can avoid the inversion shrinkage problem caused by SF to some extent, which has a more positive effect on the mechanical properties of the binder.

3.4. Phase Analysis of Binders

As can be seen in Figure 8a, no significant change in the characteristic peak of $\beta\text{-}C_2S$ in the undoped NHL was observed due to the low early hydration activity of $\beta\text{-}C_2S$, and similar results were observed after the incorporation of FA. After the incorporation of SF, the characteristic peak of $Ca(OH)_2$ showed a significant decrease due to the high early pozzolanic activity and the characteristic peak of C-S-H gel was observed at 28 d, while the characteristic peak of $\beta\text{-}C_2S$ did not show significant changes.

For the specimens doped by single B, the characteristic peaks of $Ca(OH)_2$ in the non-SCMs-dosed and FA-dosed specimens increased with the curing time, while the characteristic peaks of $\alpha'\text{-}C_2S$ showed an obvious decline, and by 28 d the characteristic peaks of $\alpha'\text{-}C_2S$ showed an extremely weak "shoulder peak", as shown in Figure 8b. This indicates a faster hydration rate of $\alpha'\text{-}C_2S$ in ion-doped binder. It should be noted that the growth of $Ca(OH)_2$ can reflect the degree of hydration of the matrix to some extent, which is applicable to both cement and NHL systems, as demonstrated in many studies [50,51]. The FA-dosed specimen (F2B1.5) showed a more significant increase in $Ca(OH)_2$ characteristic peak intensity in 28 d compared with the un-doped specimen (B1.5), and this phenomenon supports the conclusion that FA has a facilitating effect on $\alpha'\text{-}C_2S$ hydration. The $Ca(OH)_2$ peak and $\alpha'\text{-}C_2S$ peak of SF-dosed specimens both weakened along the curing time, which proved that the pozzolanic effect had a greater influence on the hydration of the SF dosed specimens. The experimental phenomenon of the coupled B/Na doped group was similar to that of the single B doped group.

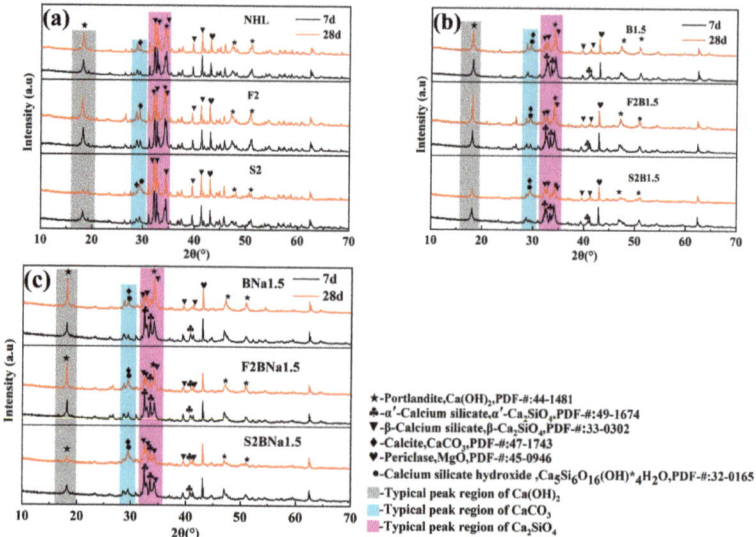

Figure 8. XRD of each sample curing for 7 d and 28 d: (**a**) blank series; (**b**) B-doped blank series; (**c**) B/Na-doped blank series.

3.5. TG Analysis

To further study the variation of the internal composition of the specimens with the degree of hydration, the variation law of chemically bound water and Ca(OH)$_2$ in the specimens at different ages was analyzed by TG. As can be seen from Figures 9 and 10 the weight loss of the specimen can be divided into four stages: −30–105 °C, 105–400 °C, 400–500 °C, and 500–800 °C. Where the weight loss below 105 °C is the evaporation of free water, 105–400 °C can be identified as C-(A)-S-H gel to remove chemically bound water, 400–500 °C is the dehydration of Ca(OH)$_2$, and 500–800 °C is the decomposition of CaCO$_3$ [52,53].

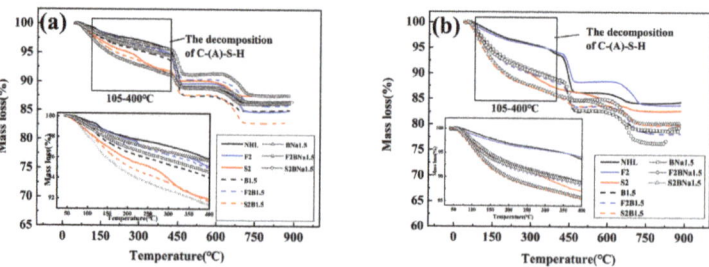

Figure 9. TG curves of the samples curing (**a**) 7 d; (**b**) 28 d.

The bound water content of all specimens increased with the extension of the curing time, and the bound water content of the doped specimens was higher than that of the blank group. At 7 d, the bound water content of single-doped B > coupled B/Na > blank group, while the strength of coupled B/Na was slightly higher than that of single-blended B at 28 d. This is consistent with the result that single-doped B specimens have higher early strength and coupled B/Na has higher late strength. This is consistent with the result that single-doped B has higher early strength and coupled B/Na has higher late strength. In the composite modified NHL coupled with SCMs, the binding water was partly derived from the hydration of C$_2$S and partly from the pozzolanic reaction, and the FA, due to its low activity and the substitution of clinker, led to a decrease in the binding water content after the FA incorporation, but in the doped group, the binding water of the FA specimens was significantly increased compared to the blank group, which again indicated that the doping had a facilitating effect on the pozzolanic activity. The higher pozzolanic activity of SF significantly enhanced the bound water in the specimens.

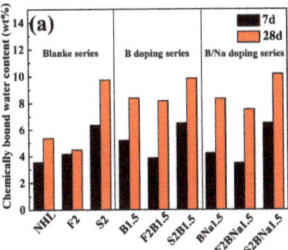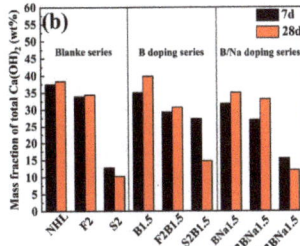

Figure 10. (a) Chemically bound water content of the samples at 7 d and 28 d; (b) total Ca(OH)$_2$ content in samples at 7 d and 28 d.

The variation of Ca(OH)$_2$ content then verifies the combined interaction of the two hydration mechanisms for the composite modified materials. For the doping samples only, the Ca(OH)$_2$ content increases with age, which corresponds to the hydration of C$_2$S, which reacts with water to produce Ca(OH)$_2$ in addition to the C-(A)-S-H gel. In contrast, for the composite modified system, the variation of Ca(OH)$_2$ content is the result of the competition between the two hydration mechanisms. For all SF-dosed samples, the Ca(OH)$_2$ content showed a decreasing trend with time, in which the Ca(OH)$_2$ content at 28 d: blank group < coupled B/Na doped < single-B doped group, which is due to the fact that the activated SiO$_2$ and Al$_2$O$_3$ would react with Ca(OH)$_2$ to form C-(A)-S-H gel. As for the doped specimens, the faster hydration of α'-C$_2$S to produce Ca(OH)$_2$ delayed the decrease of total Ca(OH)$_2$. The decrease in overall Ca(OH)$_2$ content then indicates that the strength of the pozzolanic reaction in the SF-doped composite modified NHL is greater than the hydration of C$_2$S. The Ca(OH)$_2$ content of FA-dosed composite modified NHL, on the other hand, increases slightly with the curing time, indicating that the hydration of C$_2$S is stronger than the pozzolanic reaction during its hydration.

3.6. Microstructure

The microscopic morphology of 28 d specimens was analyzed by SEM as shown in Figure 11.

For the blank specimens without the addition of SCMs, the surface is dominated by a flocculent gel that grows on the surface of the particles. This flocculent gel structure is relatively loose, as shown in Figure 11a, and this type of gel is mainly generated at the early stage of hydration of C$_2$S. After ion doping, due to the higher degree of hydration, C-(A)-S-H grows and covers the surface of C$_2$S particles, forming more widely distributed and dense gel, which is mainly needle-like with a mutual interweaving condition. The gel forms a "bridge" structure that connects particles to particles and supports the specimen, as shown in Figure 11b. For the specimens mixed with FA, two different types of gel were found. The surface of the FA particles is covered with a dense structured bulk gel, as shown in Figure 11c. Similar gel was found in Pengkun Hou's study on FA-Cement as a result of the pozzolanic reaction of FA [54]. The outside of this gel layer is covered with needle-like gel, which is generated due to the hydration of C$_2$S. For the SF-dosed specimens, the percentage of gel generated by the pozzolanic reaction is higher due to the higher pozzolanic activity of SF, as seen in Figure 11d, where the generated gel mainly appears plate-like. Furthermore, some petal-like gel was also found, and this gel was found to be similarly shaped in the morphological study by Xu, S. et al. [55] on the hydration products of NHL, which were mainly generated by the hydration of C_2S, and the morphology of the gel was related to the water-ash ratio and the degree of hydration.

Figure 11. Micromorphology of samples at 28 d: (**a**) NHL, (**b**) B1.5, (**c**) F1B1.5, and (**d**) S2B1.5.

3.7. Pore Structure

The change of porosity of the specimen also reflects the different hydration degrees, and the pore size distribution of a typical specimens cured for 28 d is shown in Figure 12.

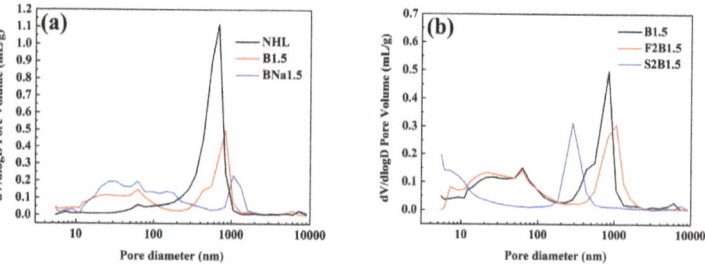

Figure 12. Pore structure distribution of typical samples cured for 28 d: (**a**) doped samples; (**b**) samples mixed with FA or SF.

The measured pores can be classified into four types according to their pore size: gel micropores (<4.5 nm); mesopores (4.5–50 nm); medium capillaries (50–100 nm); and large capillaries (>100 nm) [56]. From Figure 12a, it can be seen that the pore size of the blank NHL specimen is dominated by large capillary pores of about 700 nm, and the distribution of small pore size increases after doping, which shows that the number of large capillary pores larger than 100 nm decreases, while the number of medium-sized capillary pores within 0.01–0.1 μm increases, which indicates the increase of the dense degree of the doping matrix structure. The degree of denseness of the specimens can be judged according to the distribution of pore size: BNa1.5 > B1.5 > NHL, which is consistent with the law of compressive strength. After the addition of SCMs, there is a further reduction in the porosity of the matrix, as shown in Figure 12b. After the incorporation of FA, the number of large capillary pores was reduced due to secondary hydration and filling, while the number of mesopores around 10 nm was increased. While the samples doped with SF showed a significant decrease in the large capillary pore size, mainly concentrated around 280 nm, the content of gel micropores and mesopores increased significantly, indicating that the structural denseness was significantly improved.

4. Conclusions

The effects of ion doping and SCMs on the mechanical strength of NHL were mainly investigated. The hydration rate, mineral phase evolution pattern, and microstructure of the material under the composite action were also investigated. On this basis, the hydration mechanism under the composite action was elaborated. The conclusions were obtained as follows:

Both single-doped B and couple-doped B/Na can stabilize α'-C_2S, the stabilization effect of single-doped B on α'-C_2S is higher than that of couple-doped B/Na, and doping causes a decrease in the volume of α'-C_2S crystals. In addition, ion doping leads to the decrease of $Ca(OH)_2$ content in clinker and the increase of the amorphous phase, where the effect of complex B/Na doping is higher than that of single B doping.

Doping and the incorporation of FA and SF are able to enhance the mechanical strength of the specimens. The mechanical properties were further improved under the coupling effect. As far as doping is concerned, B doping is beneficial to the early strength of the specimens, and couple-B/Na doping is beneficial to the later strength of the specimens. After incorporation of SF, the high pozzolanic activity of SF significantly improved the early strength of the specimens, but the effect on undoped activated specimens leads to late cracking of the specimens and deterioration of the late strength. Doping can improve the degree of late cracking of the specimens. FA can effectively improve the late strength of the specimens, while for the doping group, FA incorporation specimens exhibit higher strength at an earlier period.

The composite modified lime strength is mainly related to the pozzolanic effect and α'-C_2S hydration. Doping significantly enhanced the hydration rate of the binder, with the single doped B group showing higher early hydration activity and the later hydration of the coupled B/Na, which was mainly dependent on the hydration of α'-C_2S in the specimens. The pozzolanic effect of SF is higher than the hydration of α'-C_2S in the couple-doped modification, while the pozzolanic reaction degree of FA is lower than the hydration of α'-C_2S. The hydration of the matrix can be further promoted by the synergistic effect. Under the synergistic hydration, the gel in the material coexisted with the pozzolanic-generated gel and the C_2S-generated gel, and the gel generated by the pozzolanic effect is denser in structure, which reduces the porosity of the material.

Author Contributions: Conceptualization, Y.Z. and J.L.; methodology, J.W. and D.W.; data curation, Y.Z. and C.S.; writing—original draft preparation, Y.Z. and Z.L.; writing—review and editing, Y.Z. All authors have read and agreed to the published version of the manuscript.

Funding: This research was funded by National Natural Science Foundation of China, grant number 5207020965.

Data Availability Statement: Not applicable.

Conflicts of Interest: The authors declare no conflict of interest.

References

1. Arizzi, A.; Cultrone, G. Mortars and plasters—How to characterise hydraulic mortars. *Archaeol. Anthr. Sci.* **2021**, *13*, 144. [CrossRef]
2. Elsen, J.; Mertens, G.; Snellings, R. Portland Cement and Other Calcareous Hydraulic Binders: History, Production and Mineralogy. In *Advances in the Characterization of Industrial Minerals*; Mineralogical Society of Great Britian and Ireland: Twickenham, UK, 2011; pp. 441–479.
3. Artioli, G.; Secco, M.; Addis, A. *The Vitruvian Legacy: Mortars and Binders Before and After the Roman World*; Mineralogical Society of Great Britian and Ireland: Twickenham, UK, 2019; pp. 151–202.

4. Hall, C. On the history of Portland cement after 150 years. *J. Chem. Educ.* **1976**, *53*, 222–223. [CrossRef]
5. Han, T.H.; Ponduru, S.A.; Cook, R.; Huang, J.; Sant, G.; Kumar, A. A Deep Learning Approach to Design and Discover Sustainable Cementitious Binders: Strategies to Learn From Small Databases and Develop Closed-form Analytical Models. *Front. Mater.* **2022**, *8*, 574. [CrossRef]
6. Jia, Q.Q.; Chen, W.W.; Tong, Y.M.; Guo, L.Q. Strength, hydration, and microstructure properties of calcined ginger nut and natural hydraulic lime based pastes for earthen plaster restoration. *Constr. Build. Mater.* **2022**, *323*, 126606. [CrossRef]
7. Hughes, D.C.; Jaglin, D.; Kozlowski, R.; Mucha, D. Roman cements-Belite cements calcined at low temperature. *Cem. Concr. Res.* **2009**, *39*, 77–89. [CrossRef]
8. Kalagri, A.; Miltiadou-Fezans, A.; Vintzileou, E. Design and evaluation of hydraulic lime grouts for the strengthening of stone masonry historic structures. *Mater. Struct.* **2010**, *43*, 1135–1146. [CrossRef]
9. Yang, F.W.; Zhang, B.J.; Ma, Q.L. Study of Sticky Rice-Lime Mortar Technology for the Restoration of Historical Masonry Construction. *Acc. Chem. Res.* **2010**, *43*, 936–944. [CrossRef]
10. Ruegenberg, F.; Schidlowski, M.; Bader, T.; Diekamp, A. NHL-based mortars in restoration: Frost-thaw and salt resistance testing methods towards a field related application. *Case Stud. Constr. Mater.* **2021**, *14*, e00531. [CrossRef]
11. Bauerova, P.; Reiterman, P.; Davidova, V.; Vejmelkova, E.; Storkanova, M.K.; Keppert, M. Lime Mortars with Linseed Oil: Engineering Properties and Durability. *Rev. Romana Mater.* **2021**, *51*, 239–246.
12. Silva, B.A.P.; Pinto, A.P.F.; Gomes, A. Influence of natural hydraulic lime content on the properties of aerial lime-based mortars. *Constr. Build. Mater.* **2014**, *72*, 208–218. [CrossRef]
13. Maravelaki-Kalaitzaki, P.; Bakolas, A.; Karatasios, I.; Kilikoglou, V. Hydraulic lime mortars for the restoration of historic masonry in Crete. *Cem. Concr. Res.* **2005**, *35*, 1577–1586. [CrossRef]
14. Stankeviciute, M.; Siauciunas, R.; Miachai, A. Impact of α-C2SH calcination temperature on the mineral composition and heat flow of the products. *J. Therm. Anal.* **2018**, *134*, 101–110. [CrossRef]
15. Wang, Q.; Manzano, H.; Guo, Y.; Lopez-Arbeloa, I.; Shen, X. Hydration Mechanism of Reactive and Passive Dicalcium Silicate Polymorphs from Molecular Simulations. *J. Phys. Chem. C* **2015**, *119*, 19869–19875. [CrossRef]
16. Morsli, K.; de la Torre, A.G.; Cuberos, A.J.M.; Zahir, M.; Aranda, M.A.G. Preparation and characterization of alkali-activated white belite cements. *Mater. Constr.* **2009**, *59*, 19–29.
17. Alonso, C.; Fernandez, L. Dehydration and rehydration processes of cement paste exposed to high temperature environments. *J. Mater. Sci.* **2004**, *39*, 3015–3024. [CrossRef]
18. Gawlicki, M. Effect of stabilizers on beta-C2S hydration. *Cem. Wapno Beton* **2008**, *13*, 147.
19. Huang, L.; Yang, Z. Sinterization and hydration of synthesized cement clinker doped with sulfates. *J. Therm. Anal.* **2019**, *138*, 973–981. [CrossRef]
20. Zhang, D.; Zhao, J.; Wang, D.; Xu, C.; Zhai, M.; Ma, X. Comparative study on the properties of three hydraulic lime mortar systems: Natural hydraulic lime mortar, cement-aerial lime-based mortar and slag-aerial lime-based mortar. *Constr. Build. Mater.* **2018**, *186*, 42–52. [CrossRef]
21. Uchima, J.S.; Restrepo-Baena, O.J.; Tobon, J.I. Mineralogical evolution of portland cement blended with metakaolin obtained in simultaneous calcination of kaolinitic clay and rice husk. *Constr. Build. Mater.* **2016**, *118*, 286–293. [CrossRef]
22. Wang, S.D.; Chen, C.; Gong, C.C.; Chen, Y.M.; Lu, L.C.; Cheng, X. Setting and hardening properties of alite-barium calcium sulfoaluminate cement with SCMs. *Adv. Cem. Res.* **2015**, *27*, 147–152.
23. Lin, K.-H.; Yang, C.-C. Effects of types and surface areas of activated materials on compressive strength of GGBS cement. *Mag. Concr. Res.* **2021**, *74*, 582–593. [CrossRef]
24. Cheng, Z.; An, J.; Li, F.; Lu, Y.; Li, S. Effect of fly ash cenospheres on properties of multi-walled carbon nanotubes and polyvinyl alcohol fibers reinforced geopolymer composites. *Ceram. Int.* **2022**, *48*, 18956–18971. [CrossRef]
25. Lu, D.; Shi, X.M.; Zhong, J. Interfacial nano-engineering by graphene oxide to enable better utilization of silica fume in cementitious composite. *J. Clean. Prod.* **2022**, *354*, 131381. [CrossRef]
26. Alelweet, O.; Pavia, S. Pozzolanic and hydraulic activity of bauxite for binder production. *J. Build. Eng.* **2022**, *51*, 104186. [CrossRef]
27. Grilo, J.; Faria, P.; Veiga, R.; Silva, A.S.; Silva, V.; Velosa, A. New natural hydraulic lime mortars–Physical and microstructural properties in different curing conditions. *Constr. Build. Mater.* **2014**, *54*, 378–384. [CrossRef]
28. Diouri, A.; Boukhari, A.; Aride, J.; Puertas, F.; Vazquez, T. Elaboration of alpha L′-C2S form of belite in phosphatic clinker. Study of hydraulic activity. *Mater. Constr.* **1998**, *48*, 23–32. [CrossRef]
29. Elhoweris, A.; Galan, I.; Glasser, F.P. Stabilisation of α′ dicalcium silicate in calcium sulfoaluminate clinker. *Adv. Cem. Res.* **2020**, *32*, 112–124. [CrossRef]
30. Li, C.; Wu, M.X.; Yao, W. Effect of coupled B/Na and B/Ba doping on hydraulic properties of belite-ye'elimite-ferrite cement. *Constr. Build. Mater.* **2019**, *208*, 23–35. [CrossRef]
31. Álvarez-Pinazo, G.; Cuesta, A.; García-Maté, M.; Santacruz, I.; Losilla, E.; De la Torre, A.; León-Reina, L.; Aranda, M. Rietveld quantitative phase analysis of Yeelimite-containing cements. *Cem. Concr. Res.* **2012**, *42*, 960–971. [CrossRef]
32. Segata, M.; Marinoni, N.; Galimberti, M.; Marchi, M.; Cantaluppi, M.; Pavese, A.; De la Torre, G. The effects of MgO, Na2O and SO3 on industrial clinkering process: Phase composition, polymorphism, microstructure and hydration, using a multidisciplinary approach. *Mater. Charact.* **2019**, *155*, 109809. [CrossRef]

33. Fernandes, W.; Torres, S.; Kirk, C.; Leal, A.; Filho, M.L.; Diniz, D. Incorporation of minor constituents into Portland cement tricalcium silicate: Bond valence assessment of the alite M1 polymorph crystal structure using synchrotron XRPD data. *Cem. Concr. Res.* **2020**, *136*, 106125. [CrossRef]
34. Abdelatif, Y.; Gaber, A.A.M.; Fouda, A.S.; Alsoukarry, T. Evaluation of Calcium Oxide Nanoparticles from Industrial Waste on the Performance of Hardened Cement Pastes: Physicochemical Study. *Processes* **2020**, *8*, 401. [CrossRef]
35. Gualtieri, A.F.; Viani, A.; Montanari, C. Quantitative phase analysis of hydraulic limes using the Rietveld method. *Cem. Concr. Res.* **2006**, *36*, 401–406. [CrossRef]
36. Cuesta, A.; Losilla, E.R.; Aranda, M.A.; Sanz, J.; De la Torre, G. Reactive belite stabilization mechanisms by boron-bearing dopants. *Cem. Concr. Res.* **2012**, *42*, 598–606. [CrossRef]
37. Guo, P.; Wang, B.; Bauchy, M.; Sant, G. Misfit Stresses Caused by Atomic Size Mismatch: The Origin of Doping-Induced Destabilization of Dicalcium Silicate. *Cryst. Growth Des.* **2016**, *16*, 3124–3132. [CrossRef]
38. Lai, G.C.; Nojiri, T.; Nakano, K. Studies of the Stability of Beta-Ca$_2$SiO$_4$ Doped by Minor Ions. *Cem. Concr. Res.* **1992**, *22*, 743–754. [CrossRef]
39. Haustein, E.; Kuryłowicz-Cudowska, A.; Łuczkiewicz, A.; Fudala-Książek, S.; Cieślik, B.M. Influence of Cement Replacement with Sewage Sludge Ash (SSA) on the Heat of Hydration of Cement Mortar. *Materials* **2022**, *15*, 1547. [CrossRef]
40. Wang, D.; Zhang, Q.; Feng, Y.; Chen, Q.; Xiao, C.; Li, H.; Xiang, Y.; Qi, C. Hydration and Mechanical Properties of Blended Cement with Copper Slag Pretreated by Thermochemical Modification. *Materials* **2022**, *15*, 3477. [CrossRef]
41. Amin, M.S.; Abo-El-Enein, S.A.; Rahman, A.A.; Alfalous, K.A. Artificial pozzolanic cement pastes containing burnt clay with and without silica fume Physicochemical, microstructural and thermal characteristics. *J. Anal. Calorim.* **2012**, *107*, 1105–1115. [CrossRef]
42. Zhang, D.; Zhao, J.; Wang, D.; Wang, Y.; Ma, X. Influence of pozzolanic materials on the properties of natural hydraulic lime based mortars. *Constr. Build. Mater.* **2020**, *244*, 118360. [CrossRef]
43. Yoon, S.; Choi, W.; Jeon, C. Hydration properties of mixed cement containing ground-granulated blast-furnace slag and expansive admixture. *J. Mater. Cycles Waste Manag.* **2022**, *24*, 1878–1892. [CrossRef]
44. Hasan, M.F.; Lateef, K.H. Effect of MK and SF on the concrete mechanical properties. *Mater. Today Proc.* **2021**, *42*, 2914–2919. [CrossRef]
45. Kim, M.J.; Oh, T.; Yoo, D.Y. Influence of curing conditions on the mechanical performance of ultra-high-performance strain-hardening cementitious composites. *Arch. Civ. Mech. Eng.* **2021**, *21*, 104325. [CrossRef]
46. Muthadhi, A.; Dhivya, V. Investigating Strength Properties of Geopolymer Concrete with Quarry Dust. *ACI Mater. J.* **2017**, *114*, 355–363. [CrossRef]
47. Yang, L.; Zhao, P.; Liang, C.; Chen, M.; Niu, L.; Xu, J.; Sun, D.; Lu, L. Characterization and adaptability of layered double hydroxides in cement paste. *Appl. Clay Sci.* **2021**, *211*, 106197. [CrossRef]
48. Poussardin, V.; Paris, M.; Tagnit-Hamou, A.; Deneele, D. Potential for calcination of a palygorskite-bearing argillaceous carbonate. *Appl. Clay Sci.* **2020**, *198*, 105846. [CrossRef]
49. Sharara, A.; El-Didamony, H.; Ebied, E.; El-Aleem, A. Hydration characteristics of β-C2S in the presence of some pozzolanic materials. *Cem. Concr. Res.* **1994**, *24*, 966–974. [CrossRef]
50. Lv, Y.J.; Yang, L.B.; Wang, J.L.; Zhan, B.J.; Xi, Z.M.; Qin, Y.M.; Liao, D. Performance of ultra-high-performance concrete incorporating municipal solid waste incineration fly ash. *Case Stud. Constr. Mater.* **2022**, *17*, e01155. [CrossRef]
51. Rojo-Lopez, G.; Gonzalez-Fonteboa, B.; Martinez-Abella, F.; Gonzalez-Taboada, I. Rheology, durability, and mechanical performance of sustainable self-compacting concrete with metakaolin and limestone filler. *Case Stud. Constr. Mater.* **2022**, *17*, e01143. [CrossRef]
52. Silva, A.S.; Cruz, T.; Paiva, M.J.; Candeias, A.; Adriano, P.; Schiavon, N.; Mirão, J.A.P. Mineralogical and chemical characterization of historical mortars from military fortifications in Lisbon harbour (Portugal). *Environ. Earth Sci.* **2011**, *63*, 1641–1650. [CrossRef]
53. Ubbrìaco, P.; Traini, A.; Manigrassi, D. Characterization of FDR fly ash and brick/lime mixtures. *J. Therm. Anal.* **2008**, *92*, 301–305. [CrossRef]
54. Li, P.; Ma, Z.; Li, X.; Lu, X.; Hou, P.; Du, P. Effect of Gypsum on Hydration and Hardening Properties of Alite Modified Calcium Sulfoaluminate Cement. *Materials* **2019**, *12*, 3131. [CrossRef] [PubMed]
55. Xu, S.; Wang, J.; Sun, Y. Effect of water binder ratio on the early hydration of natural hydraulic lime. *Mater. Struct.* **2015**, *48*, 3431–3441. [CrossRef]
56. Yang, Y.; Yan, Z.; Zheng, L.; Yang, S.; Su, W.; Li, B.; Ji, T. Interaction between composition and microstructure of cement paste and polymeric carbon nitride. *Constr. Build. Mater.* **2022**, *335*, 127464. [CrossRef]

Article

Effect of Different Activators on Properties of Slag-Gold Tailings-Red Mud Ternary Composite

Haonan Cui [1,*], Haili Cheng [1], Tianyong Huang [2], Feihua Yang [2], Haoxiang Lan [2] and Jvlun Li [2]

1. School of Civil Engineering, North China University of Technology, Beijing 100144, China
2. Beijing Building Materials Academy of Sciences Research, Beijing 100041, China
* Correspondence: 15689713762@163.com

Abstract: Red mud is a kind of solid waste produced in the process of aluminum extraction. Traditional methods of red mud treatment, such as open-pit accumulation and chemical recovery, are costly and environmentally hazardous. Gold tailings are industrial by-products produced in the process of gold mining and refining. In this study, NaOH, KOH, and Na_2SiO_3 were used as activators, and their effects on the properties of ternary cementitious composite containing blast furnace's slag, gold tailings, and red mud were studied with the intention of preparing a new cementitious material that is an efficient recovery and utilization of solid waste. The macroscopic mechanical properties and hydration of the ternary cementation material were studied by means of compressive strength, XRD, FT-IR, and TG/DTG. The compressive strength testing showed that the maximum strength at 28 d was 43.5 MPa. The hydration products in the ternary cementitious system were studied by SEM and EDS, and it has been demonstrated that the strength of this cement was due to the formation of Aft (AFt, also known as Ettringite, has the chemical formula $3CaO·Al_2O_3·3CaSO_4·32H_2O$. It is one of the important hydration products of cement-based cementitious materials, which can not only provide early strength for cement, but also compensate for early shrinkage of concrete.) and C-A-S-H gels. Samples activated by Na_2SiO_3 presented a most compact microstructure and the best macroscopic mechanical properties than the samples free of activator. The toxicity tests results showed that the content of heavy metal ions liberated by the cement's leaching met the standard requirements, proving that the slag-gold tailings-red mud ternary composite was environmentally friendly.

Keywords: red mud; gold tailings; geopolymer; solid waste utilization; cementitious materials

Citation: Cui, H.; Cheng, H.; Huang, T.; Yang, F.; Lan, H.; Li, J. Effect of Different Activators on Properties of Slag-Gold Tailings-Red Mud Ternary Composite. *Sustainability* 2022, 14, 13573. https://doi.org/10.3390/su142013573

Academic Editor: Ning Yuan

Received: 12 September 2022
Accepted: 12 October 2022
Published: 20 October 2022

Publisher's Note: MDPI stays neutral with regard to jurisdictional claims in published maps and institutional affiliations.

Copyright: © 2022 by the authors. Licensee MDPI, Basel, Switzerland. This article is an open access article distributed under the terms and conditions of the Creative Commons Attribution (CC BY) license (https://creativecommons.org/licenses/by/4.0/).

1. Introduction

In 2020, about 3.787 billion tons of bulk solid waste were produced in China, of which about 3.1 billion tons were hazardous, with a utilization rate estimated at only 48.67% [1]. Improving the comprehensive utilization level of hazardous solid waste is of great significance to the construction of waste free city and the recycling of renewable resources.

Red mud (RM) is one of the main solid wastes generated by the process of aluminum extraction from bauxite [1]. Producing 1 ton of alumina produces about 1 to 1.5 tons of red mud, and about 250 million tons of red mud are produced every year worldwide [2]. In 2020, China produced 106 million tons of red mud with the comprehensive utilization rate of only 7.05% [3], that is, a value far below far below the world averaged value. Indeed, 90% of the red muds produced by the global electrolytic aluminum industry is Bayer red mud [4]. This mud presents higher basicity and heavy metal content, limiting its utilization rate as resource [5]. Therefore, green development and utilization of red mud is urgent for minimizing the pollution of soil and groundwater. The authors of [6,7] investigated the effect of red mud as a mineral additive to substitute cement in the construction of concrete, that is, on its performance and application in regular manufacturing. Indeed, when the amount of red mud is 10 percent by weight, the mechanical properties of concrete are unaffected. The addition of red mud can increase the sulfate resistance of concrete.

The authors of [8] investigated the impact of red mud on the working and mechanical properties of concrete and observed that the addition of red mud had no effect on the hydration process. The mechanical strength, however, declines as the red mud component approaches 20% by weight. Similar results were reported by [9,10], respectively.

In the process of gold production and processing, considerable industrial by-products called gold tailings (GT) are produced [11]. China produced 188 million tons of gold tailings in 2020, with a comprehensive utilization rate of less than 20 percent. The main reason behind is that gold tailings after cyanidation still contain harmful substances, such as cyanide and heavy metal ions of Cd, Ni, Zn, and Pb [12,13]. Open-pit accumulation causes harmful substances release from gold tailings to the soil and change soil composition, thus damaging the environment for plant growth [14]. Therefore, a new means of disposing of gold tailings must be identified. The authors of [15] examined the effect of gold tailings substituting ordinary Portland cement (OPC), and discovered that the GT-OPC mixture met the minimum compressive strength standards for road construction. The author of [16]. studied the effect of gold tailings mixed into sulfoaluminate cement on mechanical characteristics and hydration effect. They noticed that such cement possessed superior mechanical qualities and excellent sulfate curing capacity.

In previous studies, red mud and gold tailings were traditionally added directly to cement to replace a portion of cement [17]. Although this seemed a best choice for waste solid disposal, Portland cement production was often accompanied by great carbon dioxide emissions, high energy consumption with high pollution, and waste [18]. Therefore, alternative materials with good mechanical properties that can be produced at room temperature, and without the release of secondary harmful substances, it becomes more and more attractive [19,20]. Using geopolymer materials as a substitute for traditional building materials stands for best alternative. Red mud, gold tailings, and slag with high aluminum silicate content are good raw materials for the preparation of geological polymer, that can be used, after activation, to replace traditional Portland cement concrete [21]. This not only reduces the exploitation of natural minerals but also achieves a "carbon neutral" goal [22]. Byproducts from the production of blast furnace iron are utilized to make cementitious materials because they are of high pozzolanic activity. Geopolymers can be also made from slag, and are viewed as an excellent secondary raw material [23,24].

Few studies have been done on how to prepare geopolymers using solid waste-made cementing materials composed of red mud, gold tailings, and mineral powder. Most of the research focuses on using alkali activated tailings as filling materials [16,25–27]. Furthermore, in most studies, only specific solid waste is used to develop geopolymers so that the combined use of red mud and gold tailings to produce alkali-activated cementing materials has not yet been reported [28]. This research investigates effects from NaOH, KOH, and Na_2SiO_3 (modulus = 1) as activators on the mechanics and hydration products, and evaluates microstructure of the ternary composite system, that is, slag-gold tailing-red mud made cementitious material (SGRCM). This material can not only help minimize environmental pollution from red mud and gold tailings, but also reduce the construction cost. Given that red mud, one of the raw materials, has a huge similarity to that of a sintered red brick, it may be used as a pigment to obtain red-colored mud bricks, colored pavements, colored concretes, and other decorative products. The prepared products do not need pigments or secondary coloring, directly in line with the requirements of architectural aesthetics. From a global perspective, the development of SGRCM will favor an efficient utilization of solid waste and reducing the dependence of the construction industry on traditional Portland cement.

2. Materials and Methods

2.1. Materials and Reagents

The red mud used to prepare SGRCM was provided by an aluminum plant in Shandong Province, China. Slag was supplied from a steel mill in Beijing, China. Gold tailings were sourced from a gold mining company in Yantai, China. Building plaster (BP) was purchased from the market. The reagents used in this test were: 99.9% purity sodium hydroxide, potassium hydroxide, and fine granular sodium silicate (modulus = 1). The main chemical elements of red mud, slag, and gold tailings were determined by X-ray fluorescence spectrometer (XRF), with the results shown in Table 1. Correspondingly, Table 1 also gives the loss on ignition (L.O.I.) of the raw materials. The total alkali content in the raw materials is also counted in Table 1. To more intuitively reflect the alkali concentration of the raw material, the size of the alkali concentration in the raw material is expressed by the pH value. The instrument for testing PH is PE20K laboratory PH machine, and the results are also shown at the end of Table 1. The particle size distributions of red mud, gold tailings, and slag were measured by a MS2000 laser particle size analyzer (see Figure 1) as shown in Figure 1. The size distribution of red mud was significantly smaller than that of slag and gold tailings. The radioactivity of red mud and gold tailings was determined using internal exposure index (the ratio of the specific activity of radionuclide radium-226 in the material to the limit specified in the standard) and external exposure index (the sum of the ratio of the specific activity of radionuclide radium-226, thorium 232, and potassium 40 in the material to the limit specified in the standard). The obtained results are shown in Table 2. XRD patterns of red mud, gold tailings, and slag are shown in Figure 2. XRD Quartz and feldspar ate the main mineral components of gold tailings while the red mud contains a hematite and amorphous silicoaluminate phase. As for the slag, it is mainly composed of amorphous glass phase.

Table 1. Raw materials chemical composition, loss on ignition (wt.%), and PH.

Material	SiO_2	Fe_2O_3	Al_2O_3	CaO	Na_2O	TiO_2	K_2O	SO_3	MgO	Others	Total Alkalinity	L.O.I.	PH
Red mud	22.91	22.13	18.15	3.64	9.39	3.50	0.96	0.84	0.83	0.68	7.73	5.12	10.55
Slag	28.02	0.78	16.42	37.37	-	2.51	0.50	1.80	10.04	3.76	-	0.09	10.82
Gold tailings	61.44	2.44	18.33	2.47	2.85	0.25	5.31	1.31	0.85	0.56	5.84	3.58	8.97

2.2. Test Contents and Methods

2.2.1. Sample Preparation

The cementitious material (SGRCM) is composed of red mud (10%), gold tailings (40%), slag (45%), and construction gypsum (5%). The effects of activators on the properties of the SGRCM were studied based on the mixture ratio. The amount of the added activator was set at 2 wt.%. The Chinese national standard GB/T 17671-2021 is referred to as the preparation of paste samples. The specific method is to mix the solid powder and the alkali solution mixed with different activators, pour it into a planetary mixer, and stir for 5 min. Subsequently, the slurry was uniformly cast in a steel mold with dimensions of 40 mm × 40 mm × 160 mm, cured at a temperature of 20 ± 1 °C, and had a relative humidity of 90%. The sample prepared without the activator was named SGRCM0, whereas those containing NaOH, KOH, and Na_2SiO_3 as activators were named SGRCM1, SGRCM2, and SGRCM3, respectively (see Table 3). The water-binder ratio used during the experiment (at room temperature) was set at 0.5.

2.2.2. Strength Test

Standard mortar samples with size of 40 × 40 × 160 mm were prepared according to GB/T17671-2021. The samples were cured for 3, 7, and 28 days at humidity ≥95% and temperature 20 ± 1 °C, respectively. When reaching the age, the whY-3000 pressure tester is used to test the bending strength and compressive strength of the sample. At the same time, there was a batch of 20 mm × 20 mm × 20 mm neat cement blocks, which were maintained under the same environment. The samples were taken out at 3, 7, and 28 days, respectively, and the hydration was terminated with isopropyl alcohol for microscopic analysis.

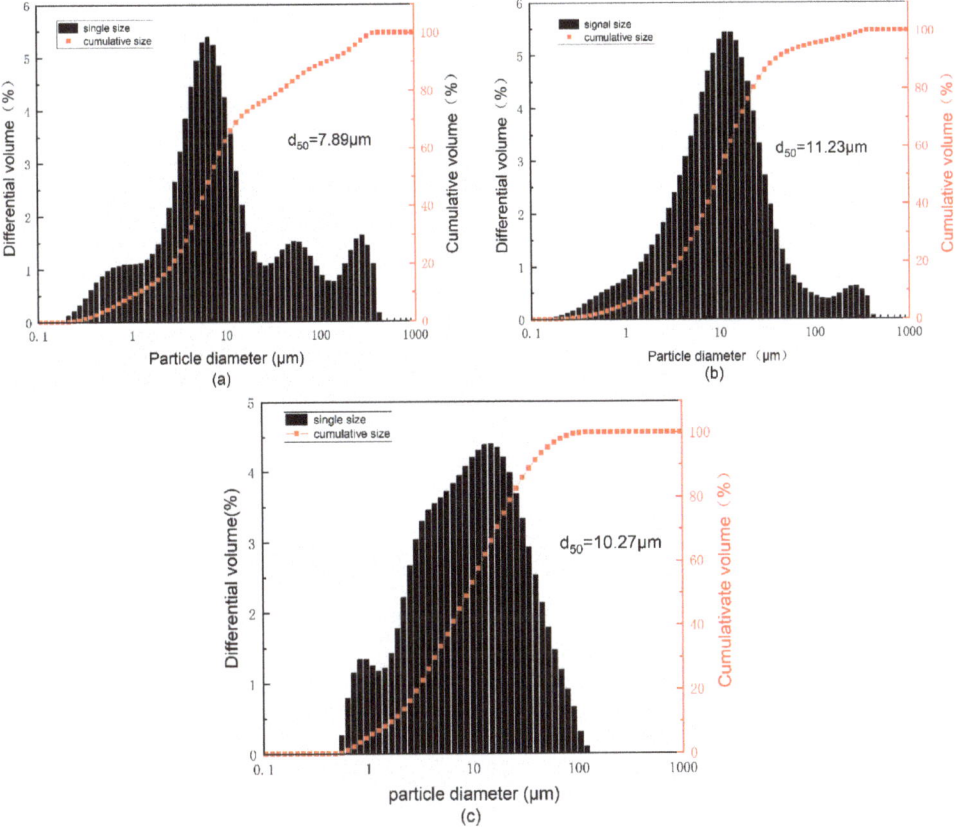

Figure 1. Particle diameter distribution of (**a**) red mud, (**b**) slag, and (**c**) gold tailings.

Table 2. Raw materials radioactivity.

Material	Internal Exposure Index	External Exposure Index
Red mud	1.83	1.78
Waste gold tailing	0.15	0.67

2.2.3. XRD Analysis

The samples were analyzed using an X-ray diffrotometer (Rigaku Corporation, Ultima IV, Tokyo, Japan), X-ray tube type: Cu target (maximum output power: 2 kW), and scanning method: THERA/2THERA Goniometer, with the samples being stationary.

2.2.4. Infrared Spectral Analysis

1 mg sample was mixed with spectrographic-graded KBr (100 mg) and homogeneously grinded. An infrared spectrometer model, iS10 (NICOLET), was used, in spectral range of 7800–350 cm^{-1}, with linear accuracy better than 0.1%.

2.2.5. Thermogravimetric Analysis

The instrument used for thermogravimetric (TG) analysis was SDT Q600 with a temperature range of 30 to 1000 °C and a heating rate of 20 °C/min in a nitrogen environment.

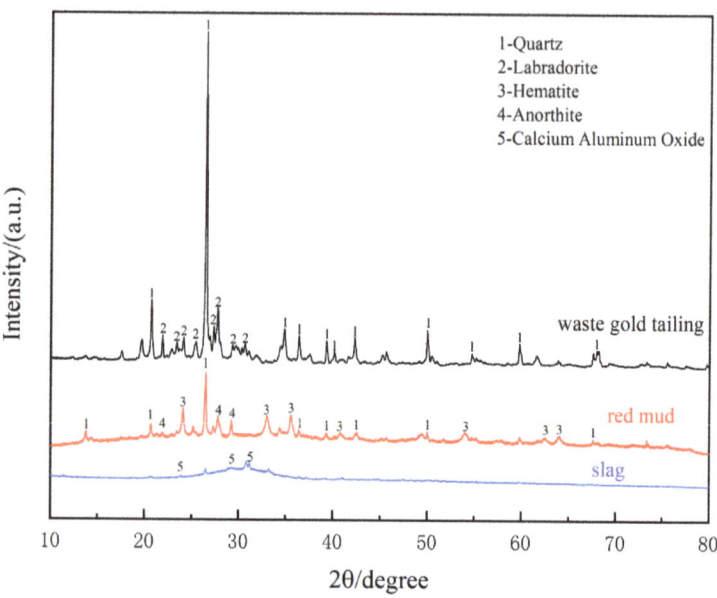

Figure 2. X-ray diffraction spectra of waste gold tailing, red mud, and slag.

Table 3. Mixing proportion of samples, wt.%.

Samples	NaOH	KOH	Na₂SiO₃	Building Plaster	Slag	GT	RM
SGRCM0	-	-	-	5	45	40	10
SGRCM1	2	-	-				
SGRCM2	-	2	-				5
SGRCM3	-	-	2				

2.2.6. Microscopic Morphology Analysis

The microscopic morphology analysis of SGRCM was taken by s-3400 N electron microscope (manufacturer: Hitachi, Marunouchi, Japan) at 20 kV. The interfaces of the different samples are knocked out, the samples are fixed to the SEM sample holder with conductive tape, and they are then sprayed with a thin layer of gold to improve electrical conductivity.

2.2.7. Leachability Test

Due to the complexity of the raw materials used to prepare SGRCM, it is necessary to evaluate the environmental impact of SGRCM. In this study, the leaching toxicity of SGRCM using different activators was conducted according to HJ/T299, 2007 standard method. The 10 g sample was crushed into particles with a diameter of smaller than 9.5 mm and mixed with 100 mL sulfuric acid/nitric acid (mass ratio M (H_2SO_4): M (HNO_3) was 2:1), with the pH set at 3.20 ± 0.05. After the sample was mixed with the leaching solution, the bottle was enclosed and fixed on the rotating oscillation device, with the speed adjusted to 30 ± 2 r/min, and the oscillation achieved at 23 ± 2 °C for 18 ± 2 h. The filter membrane was installed on the pressure filter, rinsed with dilute nitric acid, and the leaching solution was filtered with the filtrate collected by the inductively coupled plasma mass spectrometer (Model NexION 300X) in view chemical analysis.

3. Results and Discussion

3.1. Compressive Strength Analysis

The flexural strength and compressive strength of SGRCM prepared without activator (controlled samples) and samples containing 2 wt.% NaOH, KOH and Na_2SiO_3 have been measured according to the specification GB/T17671-2021 as shown by the results depicted in Figure 3. The addition of NaOH and KOH decreased the samples strength compared to that of the control sample. This is due to the fact that the red mud, gold tailings, and slag are alkaline materials, so they had brought too much OH^- chemical species. This resulted in increased internal alkalinity and in a speeded precipitation of hydration products on the surface of glassy particles inhibiting their hydration reaction. The addition of Na_2SiO_3 did not contribute to sample early strength and after 7 d, the compressive strength was improved with the increase of curing time. At 7 d, the compressive strength was 30.41% higher than the value given the control sample and has reached 34.43 MPa. This revealed that the sample activated using Na_2SiO_3 displayed a better mechanical performance when compared to other samples. Moreover, Na_2SiO_3 itself has adhesive properties capable of improving the strength of the sample at an early stage.

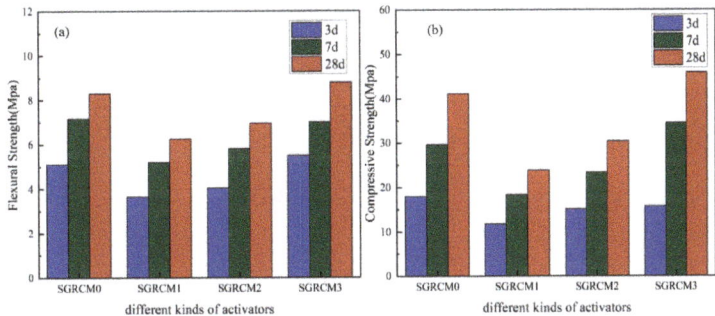

Figure 3. Effects of different activators on flexural strength (**a**) and compressive strength (**b**).

3.2. XRD Analysis

To determine the mineral phase's composition of hydration products, XRD analysis was performed on SGRCM blocks prepared with different activators, and were analyzed by XRD after different curing ages (3, 7, 28 d), as shown in Figure 4. Figure 4A presents the X-ray diffraction patterns of SGRCM prepared without the activator after 3, 7, and 28 days. As for Figure 4B–D show the XRD patterns of SGRCM activated by NaOH, KOH, and Na_2SiO_3 (modulus = 1). Figure 4A–D reveals that after 28 d, ettringite (AFt), mullite, and anorthite ($2CaO \cdot Al_2O_3 \cdot SiO_2$) are the primary hydration products [29]. Ettringite consists of crystallized hydrated calcium sulfoaluminate produced by the hydration reaction involving calcium aluminates and sulfate ion. This indicates that the hydration products formed by the active aluminates in the raw materials were the main source of strength shown by SGRCM.

Residual quartz was also found in the raw material. It may cause micro aggregate reaction, that is, a positive effect on improving strength. In addition, the diffraction peaks of ettringite and mullite were seen at 3 d in samples prepared without an activator, but those peaks are much weaker in samples prepared with NaOH and KOH as activator. This indicates that the addition of NaOH and KOH has inhibited the formation of ettringite. The alkalinity provided by red mud, gold tailings, and ore powder (red mud pH = 10.55, gold tailings pH = 8.79, slag pH = 10.82) was sufficient for converting glassy phases in ore powder into C-A-S-H. However, after the addition of alkali, pH was so high that the C-A-S-H gels generated quickly precipitated on the glass phase surface [30,31]. Consequently, the glass phase failed to contact the alkali solution, so the hydration reaction is inhibited [32,33].

On the contrary, the addition of Na$_2$SiO$_3$ resulted in a slight pH change, therefore the introduced silicon promoted the hydration reaction [34].

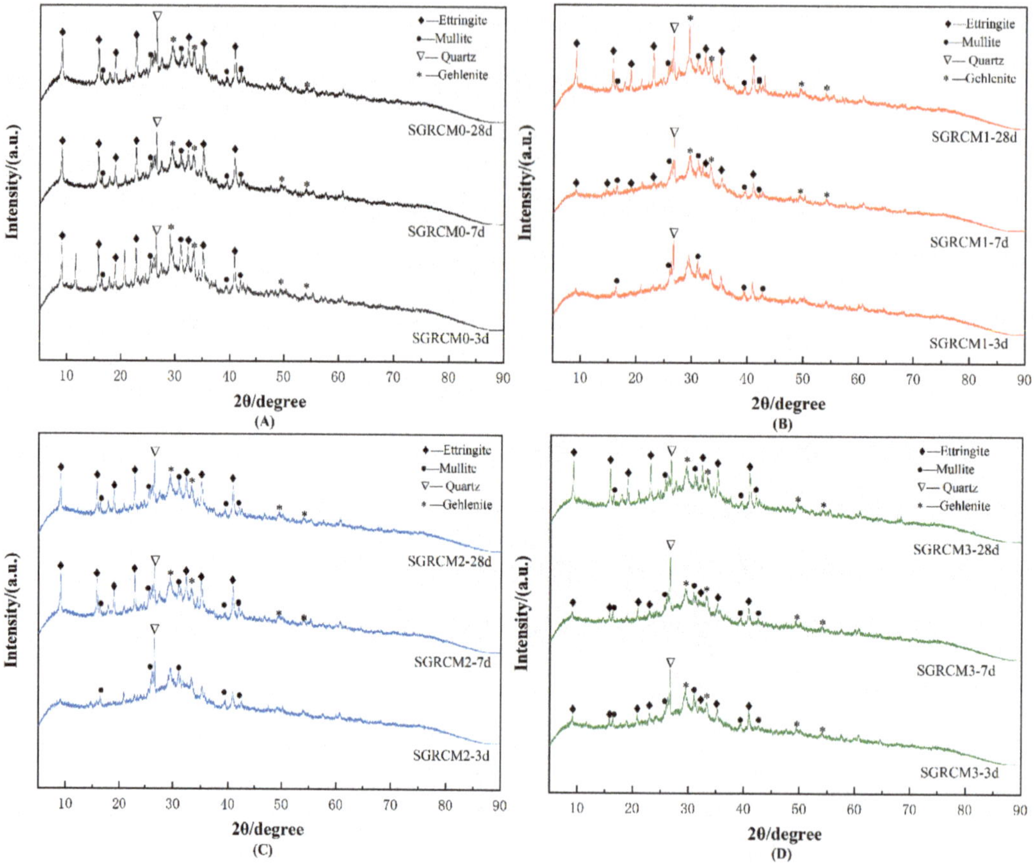

Figure 4. XRD spectrum of the SGRCM cured for 3 d/7 d/28 d with different activators. (**A**) shows the XRD plots of SGRCM at 3, 7, and 28 d without the addition of an activator. (**B–D**) are the XRD plots at 3, 7, and 28 d when NaOH, KOH, and Na$_2$SiO$_3$ were added as the activator.

3.3. FT-IR Spectroscopy Analysis

Figure 5 depicts the FT-IR spectra of SGRCM at 3 d and 28 d. It can be seen that the FT-IR spectra of samples prepared with or without activators look basically the same. The absorption peak seen at 3430 cm^{-1} stands for asymmetric stretching vibration of hydroxyl (X-OH) [35]. As for the absorption peak seen at 1630 cm^{-1}, it is associated with the bending vibration of H-O-H in C-A-S-H [27]. Under the combined activation of red mud, gold tailings and gypsum, SGRCM produced numerous hydrates (products containing water). In Figure 4, there is a weak absorption peak at 3150 cm^{-1}, which is caused by the stretching vibration of OH$^-$. The strong absorption peak at 1090 cm^{-1} is caused by the asymmetric stretching vibration of SO$_4^{2-}$ in AFt. This reveals that ettringite can be generated in both the early stage and the late stage regardless of the activator used. It is clear that the ettringite crystals are involved in the early strength of SGRCM. The results are consistent with those from XRD and SEM analyses. In addition, the absorption peak at 1440 cm^{-1} is caused by CO$_3^{2-}$ antisymmetric stretching vibration due to carbonization of samples exposed to air. There are also two absorption peaks between 400 cm^{-1} and 1000 cm^{-1}, associated to the tensile and angular bending vibration of Si-O-Al and Si-O-Si bonds [36]. This indicates the presence of silicaluminate groups in the geopolymer structure, similar to the hydrated calcium silicaluminate (C-A-S-H) formed during cement hydration. The FTIR spectra of different SGRCM activated using different activators did not show any obvious difference. This reveals the fact that different activators do not reduce the formation of aluminosilicate gel, but only change the intensity of the band associated with the Si and Al bonds.

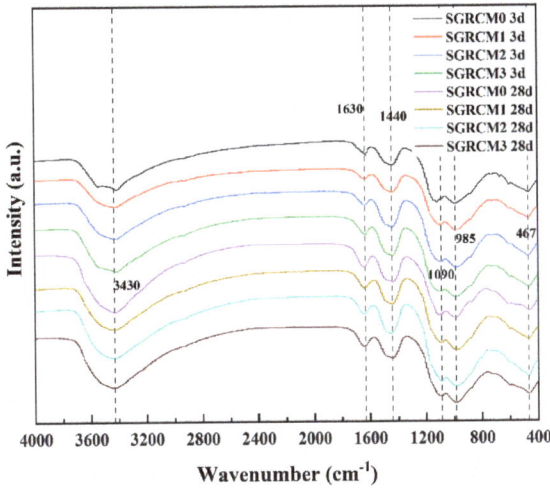

Figure 5. Geopolymer patterns produced by FTIR.

3.4. Thermogravimetric Analysis

The thermogravimetric and differential heat curves of SGRCM hardened slurry at 1 d and 28 d are shown in Figure 6. As can be seen from the figure, there are three weight loss intervals for all samples, and 28 d samples have greater weight loss in each interval than the 1 d samples. The first weight loss interval was 50–250 °C, with two endothermal peaks at 100 °C and 150 °C on the DTG curve, respectively. The peak at 100 °C was caused by the loss of crystallization water from hydration product AFt and the loss of interlayer water of C-A-S-H gel [11]. The peak at 150 °C has probably been caused by the dehydration of gypsum into hemihydrate gypsum. By comparing the DTG curves at 28 d and 3 d, it can be seen that the two peaks at 100 °C and 150 °C at 3 d emerge as the ones seen at about 100 °C after 28 d. The peak become wider and sharper obviously. given that the content of C-A-S-H gel and AFt has increased with the extension of curing age [37]. After 28 d, the peak at 150 °C disappeared, showing that gypsum was gradually and completely consumed with the increase of hydration age. The peak at 450 °C in the DTG curves of samples cured for 1 d and 28 d demonstrates the presence of Ca (OH)$_2$. of which the peak weakened after 28 d due to the continuous consumption of Ca (OH)$_2$ by the pozzolanic reactions. For samples cured for 3 d and 28 d, a peak was also seen at 600–800 °C are also found. This peak stands for the CaCO$_3$ produced through carbonation of Ca (OH)$_2$ in the presence of CO$_2$ contained in air. The above reveals that SGRCM has directly captured some carbon dioxide when exposed to air.

Figure 6c is the TG curve of SGRCM after 28 d hydration. As can be seen, after 28 d, the weight loss of SGRCM3 during the whole heating regime is the highest, indicating that more hydration products are formed in SGRCM3 compared with other groups. Figure 6d is the DTG curve of SGRCM after 28 d hydration. As can be seen, the endothermic peak of SGRCM3 at 100 °C is widest and sharpest. This suggests considering that SGRCM3 was the highest content of C-A-S-H and AFt under the activation of Na$_2$SiO$_3$, and correspondingly, the least Ca (OH)$_2$ was dehydrated at 450 °C. The obtained results thus indicate that in the pozzolanic reaction activated by Na$_2$SiO$_3$, more Ca(OH)$_2$ crystals were converted to C-A-S-H gels [38,39]. Combined with results from microscopic analysis, one can see that C-A-S-H gel in SGRCM3 has filled the voids between AFt crystals, resulting in improved macroscopic mechanical properties shown by SGRCM3.

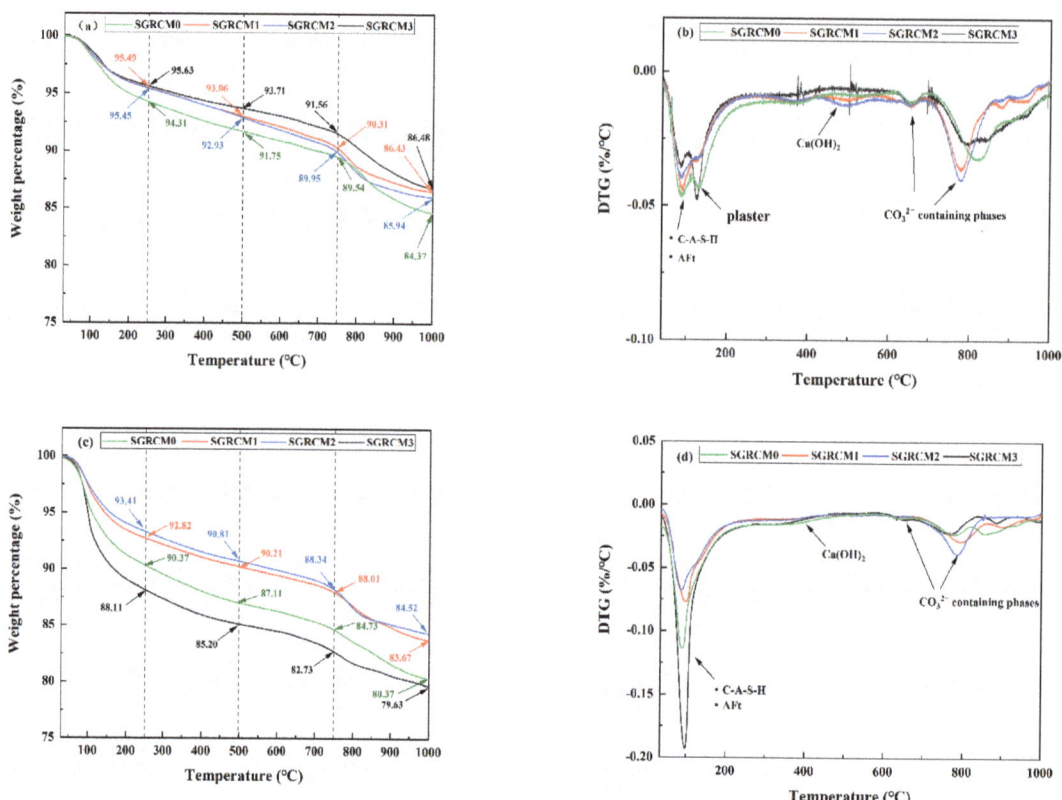

Figure 6. TGA/DTG curves of SGRCM under different activators for (**a**,**b**) 1 d (**c**,**d**) 28 d.

3.5. SEM and EDS Analysis

Figures 7 and 8 depict the microscopic morphology of the hydration products of SGRCM0, SGRCM1, and SGRCM3 paste samples after 1 d and 28 d of curing. The amorphous network of the SGRCM0 paste sample was identified as a C-A-S-H gel using EDS analysis (Figure 7 region A). This network connected irregular hydration products of differing sizes in a loose manner [40]. Even though some unreacted particles were identified in local regions, the C-A-S-H gel covers the surface of these particles and fills the pores between particles of varying sizes forming a network structure that enhanced compressive strength. Needle-rod ettringite was detected in the C-A-S-H gel of the Na$_2$SiO$_3$-activated SGRCM3 sample.

The elemental composition of hydration products was determined using energy dispersive X-ray diffraction (EDS) analysis (Figure 7, areas A, B, and C). A single day of hydration leads to the formation of amorphous C-A-S-H gel in SGRCM0, SGRCM1, and SGRCM3 samples, analyzed by EDS. As the hydration process advanced, the Al^{3+}, K$^+$, and Na$^+$ in the sample were fixed in the hydration products. According to previous research, Al^{3+} might substitute Si^{4+} to bridge tetrahedral sites, generating a long chain structure of aluminum-containing C-A-S-H gel conducive to the development of compressive strength.

The net negative charge caused by substituting tetrahedral aluminum for silicon was balanced by other cations, such as K^+ and Na^+.

By comparison, C-A-S-H and ettringite were both produced under the activation of NaOH and Na_2SiO_3. Ettringite formed in Na_2SiO_3-activated samples was interlaced with C-A-S-H, and C-A-S-H also filled the ettringite's internal pores. In contrast, only ettringite was produced on the surface of SGRCM1 when the activation was done using NaOH, NaOH, and ettringite was neither dense nor compact. This explains the reason why its strength was very small.

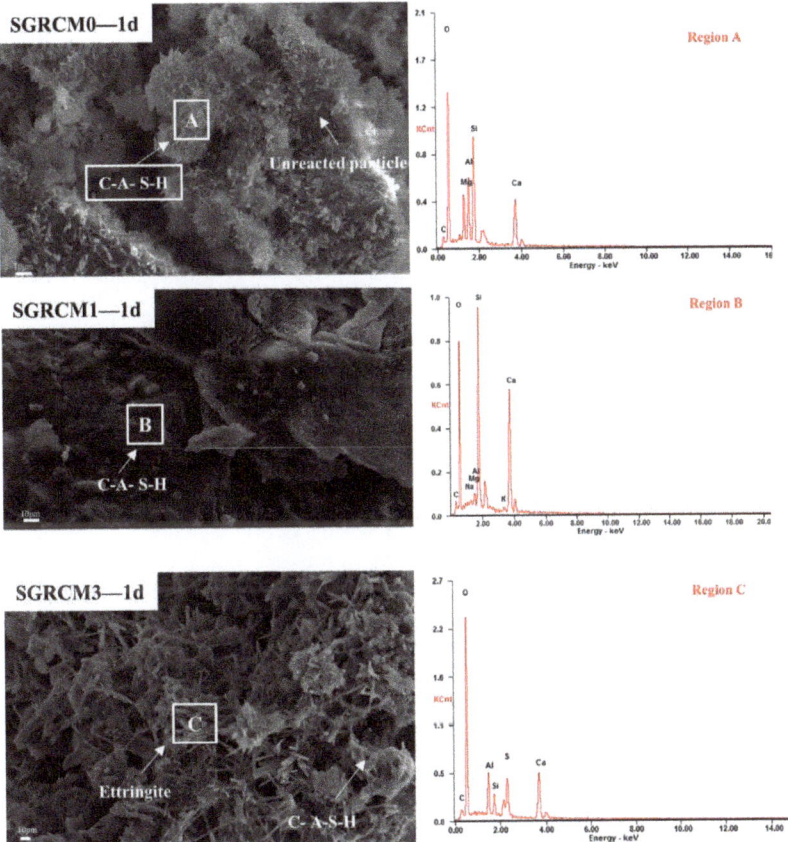

Figure 7. SEM images and EDS analysis of SGRCM0, SGRCM1, and SGRCM3 samples at 1 d.

As can be seen from SGRCM3-28d in Figure 7, the microstructure of the SGRCM slurry is dense, a thick amorphous C-A-S-H gel is observed, and the AFt is covered by the C-A-S-H gel. Rod-shaped AFt crystals and C-A-S-H gels are distributed in the holes and pits, which gradually grow and fuse, and the overall porosity of the material is greatly reduced. Therefore, the compactness of SGRCM3 is relatively high, and the macroscopic performance is high in mechanical properties.

The results of SEM and EDS analyses demonstrate once more that the principal hydration products of red mud, gold tailings, and ore powder were compound of amorphous C-A-S-H gel and fibrous ettringite crystals. As an activator, silicate enhanced the pozzolana activity of red mud and gold tailings, and promote the development of early strength. The activation effect of silicate was significantly superior to that of alkali and consistent with some of their research [41,42]; thus, the geopolymer samples prepared in the present study were close to the typical normal geopolymer samples.

3.6. Ion Leaching Analysis

Raw material of slag, red mud, gold tailings, and SGRCM samples were leached using the sulfuric acid and nitric acid methods. Inductively coupled plasma mass spectrometry (ICP-MS) enabled ascertaining the toxicity of solutions from the leaching. Table 4 displays the concentrations and critical values of heavy metals in the leaching solution. After 28 dof curing, the concentration of heavy metal ions in SGRCM material revealed to comply with the HJ/T 299-2007 standard. As can be seen from Table 4, the concentration of all heavy metal ions in the leaching solution of sample SGRCM3 at 28 d were less than that at 1 d, showing the capacity of SGRCM3 to successfully solidify heavy metal ions. Crystals can solidify heavy metal ions because heavy metal ions can replace original ions in the crystal or enter voids in the crystal lattice [43]. The XRD pattern in Figure 4 has revealed that calcium-aluminite phases were present in hydration products. Figure 9 depicts the structural mechanism of anorthite formation. Typically, the basic structural unit of anorthite crystals is the tetrahedron, which is generated by the connection of four tetrahedrons and developed into three-dimensional space to form the basic framework. During hydration, Si-O tetrahedron is reorganized, and a portion of Si^{4+} in $[SiO_4]$ tetrahedron can be replaced by Al^{3+} to produce $[AlO_4]$ tetrahedron, producing a three-dimensional network structure consisting of $[SiO_4]$ and $[AlO_4]$ tetrahedron. The structure of this Al-Si frame is negatively charged. Consequently, the presence of heavy metal cations in the void of the frame contributes to the system's charge balance [44]. When heavy metal ions are included, due to the high bond strength of the calcium feldspar system, crystal nuclei are formed from the surrounding crystalline phases, and other heavy metal ions may be attracted to partially replace Si^{4+} or Al^{3+} ions to form similar $[XO_4]$ tetrahedrons; this is a process that solidifies the heavy metal elements in the crystal lattice. Indeed, heavy metal ions can also replace calcium cations in anorthite crystal gaps [45]. Ca^{2+} ions in the anorthite phase can be substituted by Cu ions and Mg ions when heavy metal ions share the same ionic characteristics as calcium cations (e.g., the same positive bivalent charge) [46]. The above analysis demonstrates that heavy metal elements can be efficiently solidified in the SGRCM samples via two types of mechanisms: one is to substitute Si^{4+} or Al^{3+} ions in $[SiO_4]$ or $[AlO_4]$ tetrahedra to form $[XO_4]$ tetrahedra (X is a heavy metal ion). The alternative is to replace Ca^{2+} in the interlayer and enter the network gap of the skeleton structure in order to counteract the negative charge. The cementitious materials prepared from red mud, waste gold tailings, and slag are therefore environmentally acceptable.

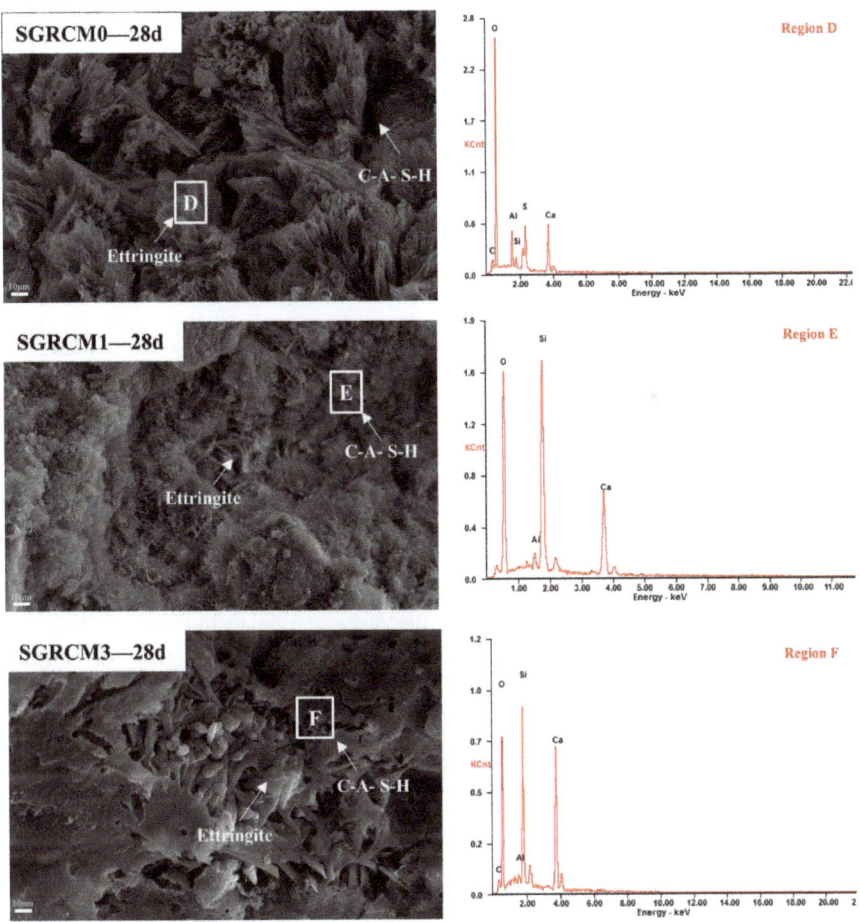

Figure 8. SEM images and EDS analysis of SGRCM0, SGRCM1, and SGRCM3 samples at 28 d.

Table 4. Leaching concentration of heavy metals in SGRCM with their critical limits.

Sample	Leaching Concentration of Heavy Metals				
	Al (mg/L)	As (mg/L)	Cu (mg/L)	Mo (mg/L)	Cr (mg/L)
SL	3.388	0	0	0.009	0
GT	0.053	0.080	0.005	0.016	0
RM	9.500	0.927	0.030	0.100	0.156
SGRCM0-1d	1.065	0.011	0.005	0.107	0.017
SGRCM1-1d	1.136	0.014	0.018	0.125	0.019
SGRCM2-1d	1.548	0.017	0.010	0.121	0.020
SGRCM3-1d	3.532	0.007	0.013	0.107	0.011
SGRCM0-28d	0.809	0.008	0.002	0.060	0.010
SGRCM1-28d	1.598	0.011	0.017	0.102	0.009
SGRCM2-28d	2.770	0.010	0.012	0.126	0.009
SGRCM3-28d	0.708	0	0.001	0.957	0.007
GB5085.3-2007 critical limits	100	5	100	-	1

In SGRCM, the ion concentrations of Cd, Ag, Mn, Pb and Zn are all zero, so they are not listed.

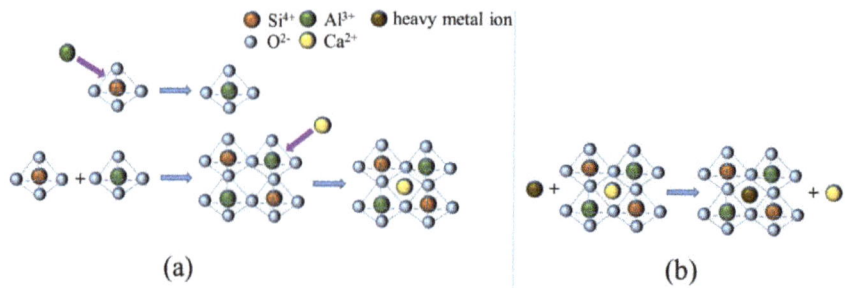

Figure 9. (**a**) The structural mechanism of anorthite formation, (**b**) the replacement of Ca by heavy metal ions.

4. Discussion and Conclusions

This paper attempts to use solid waste (red mud, gold tailings, and slag) instead of cement as a cementitious material in construction, providing a reference for future research on more green and environmentally friendly cementitious materials. In this study, the basic engineering indexes and environmental safety evaluation indexes of the specimens excited by different kinds of activators were tested. Meanwhile, the hydration products were characterized by XRD and FTIR. The main conclusions of this study are as follows:

1. The mass ratio of M (red mud) to M (gold tailings) to M (slag) to M (gypsum) in this study is 10:40:45:5. After curing at room temperature for 28 d, the compressive strength of SRCM3 reached 43.7 MPa.
2. According to XRD analyses, the hydration products of SGRCM3 are mainly composed of C-A-S-H and AFt, and this was in line with data from SEM and EDS analyses. Those hydration products are believed to have promoted the high compressive strength obtained.
3. The leaching concentrations of heavy metals in the samples were below the detection level, suggesting that the concentration of heavy metal ions in SGRCM were below the permitted limit for the environment. The obtained results indicate that the red mud-gold, tailing-slag, ternary composite cementitious material can consolidate heavy metal ions in a red mud.
4. The findings of the SRCM study indicate that the red-mud-gold-tailing-slag composite can be employed as a substitute for cement in industry when it is activated by Na_2SiO_3 (modulus = 1).

Author Contributions: Conceptualization, H.C. (Haonan Cui) and J.L.; methodology, H.C. (Haonan Cui) and T.H.; software, H.C. (Haonan Cui) and H.L.; validation, F.Y., H.C. (Haili Cheng) and H.L.; formal analysis, H.C. (Haili Cheng); investigation, J.L.; resources, T.H.; data curation, H.C. (Haonan Cui); writing—original draft preparation, H.C. (Haonan Cui); writing—review and editing, F.Y. and H.C. (Haili Cheng); visualization, H.C. (Haonan Cui); supervision, F.Y.; project administration, T.H. and F.Y.; funding acquisition, H.C. (Haili Cheng) All authors have read and agreed to the published version of the manuscript.

Funding: This research was funded by Beijing Building Materials Academy of Sciences Research, grant number 2021YFC1910605 and the APC was funded by School of Civil Engineering, North China University of Technology.

Institutional Review Board Statement: Not applicable.

Informed Consent Statement: Not applicable.

Data Availability Statement: Not applicable.

Acknowledgments: The authors are grateful for the support provided by the National Key R&D Program of China (2021YFC1910605), and Beijing Building Materials Academy of Sciences Research State key laboratory of solid waste resource utilization and energy saving building materials for the kind assistance in laboratory testing.

Conflicts of Interest: All authors disclosed no relevant relationships. The authors declared no potential conflicts of interest with respect to the research, authorship, and/or publication of this article.

References

1. Qi, Y. The neutralization and recycling of red mud—A review. *J. Phys. Conf. Ser.* **2021**, *1759*, 012004. [CrossRef]
2. Borra, C.R.; Blanpain, B.; Pontikes, Y.; Binnemans, K.; Van Gerven, T. Recovery of Rare Earths and Other Valuable Metals from Bauxite Residue (Red Mud): A Review. *J. Sustain. Metall.* **2016**, *2*, 365–386. [CrossRef]
3. Li, H.; Ye, H.; Zhu, J.; Liu, J. Research status of Bayer process red mud dealkalization technology. *J. Shandong Univ. Technol.* **2021**, *35*, 65–69.
4. Gräfe, M.; Power, G.; Klauber, C. Bauxite residue issues: III. Alkalinity and associated chemistry. *Hydrometallurgy* **2011**, *108*, 60–79. [CrossRef]
5. Ye, J.; Hu, A.; Ren, G.; Zhou, T.; Zhang, G.; Zhou, S. Red mud enhances methanogenesis with the simultaneous improvement of hydrolysis-acidification and electrical conductivity. *Bioresour. Technol.* **2018**, *247*, 131–137. [CrossRef] [PubMed]
6. Gelencser, A.; Kovats, N.; Turoczi, B.; Rostasi, A.; Hoffer, A.; Imre, K.; Nyiro-Kosa, I.; Csakberenyi-Malasics, D.; Toth, A.; Czitrovszky, A.; et al. The red mud accident in Ajka (Hungary): Characterization and potential health effects of fugitive dust. *Environ. Sci. Technol.* **2011**, *45*, 1608–1615. [CrossRef]
7. Ghalehnovi, M.; Roshan, N.; Hakak, E.; Shamsabadi, E.A.; de Brito, J. Effect of red mud (bauxite residue) as cement replacement on the properties of self-compacting concrete incorporating various fillers. *J. Clean. Prod.* **2019**, *240*, 118213. [CrossRef]
8. Senff, L.; Hotza, D.; Labrincha, J.A. Effect of red mud addition on the rheological behaviour and on hardened state characteristics of cement mortars. *Constr. Build. Mater.* **2011**, *25*, 163–170. [CrossRef]
9. Singh, B.; Ishwarya, G.; Gupta, M.; Bhattacharyya, S.K. Geopolymer concrete: A review of some recent developments. *Constr. Build. Mater.* **2015**, *85*, 78–90. [CrossRef]
10. Zhang, N.; Sun, H.; Liu, X.; Zhang, J. Early-age characteristics of red mud–coal gangue cementitious material. *J. Hazard. Mater.* **2009**, *167*, 927–932. [CrossRef]
11. Yao, G.; Liu, Q.; Wang, J.; Wu, P.; Lyu, X. Effect of mechanical grinding on pozzolanic activity and hydration properties of siliceous gold ore tailings. *J. Clean. Prod.* **2019**, *217*, 12–21. [CrossRef]
12. Dehghani, A.; Mostad-Rahimi, M.; Mojtahedzadeh, S.; Gharibi, K. Recovery of gold from the Mouteh Gold Mine tailings dam. *J. South. Afr. Inst. Min. Metall.* **2009**, *109*, 417–421.
13. Ndivhudzannyi, R.; Dacosta, F.; Gumbo, J. *Environmental Risk Assessment and Risk Management Strategies for Dysfunctional New Union Gold Mine in Malamulele, Limpopo, South Africa*; China University of Mining and Technology Press: Xuzhou, China, 2014.
14. Maroušek, J.; Stehel, V.; Vochozka, M.; Kolář, L.; Maroušková, A.; Strunecký, O.; Peterka, J.; Kopecký, M.; Shreedhar, S. Ferrous sludge from water clarification: Changes in waste management practices advisable. *J. Clean. Prod.* **2019**, *218*, 459–464. [CrossRef]
15. Rachman, R.; Bahri, A.; Trihadiningrum, Y. Stabilization and solidification of tailings from a traditional gold mine using Portland cement. *Environ. Eng. Res.* **2018**, *23*, 189–194. [CrossRef]
16. Kiventerä, J.; Golek, L.; Yliniemi, J.; Ferreira, V.; Deja, J.; Illikainen, M. Utilization of sulphidic tailings from gold mine as a raw material in geopolymerization. *Int. J. Miner. Process.* **2016**, *149*, 104–110. [CrossRef]
17. Consoli, N.C.; Nierwinski, H.P.; Peccin da Silva, A.; Sosnoski, J. Durability and strength of fiber-reinforced compacted gold tailings-cement blends. *Geotext. Geomembr.* **2017**, *45*, 98–102. [CrossRef]
18. Wu, Y.; Lu, B.; Bai, T.; Wang, H.; Du, F.; Zhang, Y.; Cai, L.; Jiang, C.; Wang, W. Geopolymer, green alkali activated cementitious material: Synthesis, applications and challenges. *Constr. Build. Mater.* **2019**, *224*, 930–949. [CrossRef]
19. Miranda, T.; Leitão, D.; Oliveira, J.; Corrêa-Silva, M.; Araújo, N.; Coelho, J.; Fernández-Jiménez, A.; Cristelo, N. Application of alkali-activated industrial wastes for the stabilisation of a full-scale (sub)base layer. *J. Clean. Prod.* **2020**, *242*, 118427. [CrossRef]
20. Tonini de Araújo, M.; Tonatto Ferrazzo, S.; Jordi Bruschi, G.J.; Consoli, N.C. Mechanical and Environmental Performance of Eggshell Lime for Expansive Soils Improvement. *Transp. Geotech.* **2021**, *31*, 100681. [CrossRef]
21. Acordi, J.; Luza, A.; Fabris, D.C.N.; Raupp-Pereira, F.; De Noni, A., Jr.; Montedo, O.R.K. New waste-based supplementary cementitious materials: Mortars and concrete formulations. *Constr. Build. Mater.* **2020**, *240*, 117877. [CrossRef]
22. Sithole, T.; Mashifana, T. Geosynthesis of building and construction materials through alkaline activation of granulated blast furnace slag-NC-ND license. *Constr. Build. Mater.* **2020**, *264*, 120712. [CrossRef]
23. Singh, S.; Aswath, M.U.; Ranganath, R.V. Effect of mechanical activation of red mud on the strength of geopolymer binder. *Constr. Build. Mater.* **2018**, *177*, 91–101. [CrossRef]
24. Wang, J.; Lyu, X.; Wang, L.; Cao, X.; Liu, Q.; Zang, H. Influence of the combination of calcium oxide and sodium carbonate on the hydration reactivity of alkali-activated slag binders. *J. Clean. Prod.* **2018**, *171*, 622–629. [CrossRef]
25. Ince, C. Reusing gold-mine tailings in cement mortars: Mechanical properties and socio-economic developments for the Lefke-Xeros area of Cyprus. *J. Clean. Prod.* **2019**, *238*, 117871. [CrossRef]

26. Jiang, H.; Qi, Z.; Yilmaz, E.; Han, J.; Qiu, J.; Dong, C. Effectiveness of alkali-activated slag as alternative binder on workability and early age compressive strength of cemented paste backfills. *Constr. Build. Mater.* **2019**, *218*, 689–700. [CrossRef]
27. Zhang, X.-F.; Ni, W.; Wu, J.-Y.; Zhu, L.-P. Hydration mechanism of a cementitious material prepared with Si-Mn slag. *Int. J. Miner. Metall. Mater.* **2011**, *18*, 234–239. [CrossRef]
28. Adesina, A.; Das, S. Influence of glass powder on the durability properties of engineered cementitious composites. *Constr. Build. Mater.* **2020**, *242*, 118199. [CrossRef]
29. Liu, X.; Zhang, N.; Sun, H.; Zhang, J.; Li, L. Structural investigation relating to the cementitious activity of bauxite residue—Red mud. *Cem. Concr. Res.* **2011**, *41*, 847–853. [CrossRef]
30. Cheah, C.B.; Tan, L.E.; Ramli, M. The engineering properties and microstructure of sodium carbonate activated fly ash/slag blended mortars with silica fume. *Compos. Part B Eng.* **2019**, *160*, 558–572. [CrossRef]
31. Somna, K.; Jaturapitakkul, C.; Kajitvichyanukul, P.; Chindaprasirt, P. NaOH-activated ground fly ash geopolymer cured at ambient temperature. *Fuel* **2011**, *90*, 2118–2124. [CrossRef]
32. Alonso, S.; Palomo, A. Alkaline activation of metakaolin and calcium hydroxide mixtures: Influence of temperature, activator concentration and solids ratio. *Mater. Lett.* **2001**, *47*, 55–62. [CrossRef]
33. Khale, D.; Chaudhary, R. Mechanism of geopolymerization and factors influencing its development: A review. *J. Mater. Sci.* **2007**, *42*, 729–746. [CrossRef]
34. Singh, P.S.; Bastow, T.; Trigg, M. Structural studies of geopolymers by 29Si and 27Al MAS-NMR. *J. Mater. Sci.* **2005**, *40*, 3951–3961. [CrossRef]
35. Mladenovič, A.; Mirtič, B.; Meden, A.; Zalar Serjun, V. Calcium aluminate rich secondary stainless steel slag as a supplementary cementitious material. *Constr. Build. Mater.* **2016**, *116*, 216–225. [CrossRef]
36. Chen, H.; Yuan, H.; Mao, L.; Hashmi, M.Z.; Xu, F.; Tang, X. Stabilization/solidification of chromium-bearing electroplating sludge with alkali-activated slag binders. *Chemosphere* **2020**, *240*, 124885. [CrossRef]
37. Collier, N.C. Transition and Decomposition Temperatures of Cement Phases—A Collection of Thermal Analysis Data. *Ceram. Silik.* **2016**, *60*, 338–343. [CrossRef]
38. Rashad, A.M.; Zeedan, S.R.; Hassan, A.A. Influence of the activator concentration of sodium silicate on the thermal properties of alkali-activated slag pastes. *Constr. Build. Mater.* **2016**, *102*, 811–820. [CrossRef]
39. Wan, Q.; Rao, F.; Song, S.; García, R.E.; Estrella, R.M.; Patiño, C.L.; Zhang, Y. Geopolymerization reaction, microstructure and simulation of metakaolin-based geopolymers at extended Si/Al ratios. *Cem. Concr. Compos.* **2017**, *79*, 45–52. [CrossRef]
40. Li, H.; Guan, X.; Zhang, X.; Ge, P.; Hu, X.; Zou, D. Influence of superfine ettringite on the properties of sulphoaluminate cement-based grouting materials. *Constr. Build. Mater.* **2018**, *166*, 723–731. [CrossRef]
41. Guo, X.; Zhang, L.; Huang, J.; Shi, H. Detoxification and solidification of heavy metal of chromium using fly ash-based geopolymer with chemical agents. *Constr. Build. Mater.* **2017**, *151*, 394–404. [CrossRef]
42. Nath, S.K.; Kumar, S. Reaction kinetics, microstructure and strength behavior of alkali activated silico-manganese (SiMn) slag –Fly ash blends. *Constr. Build. Mater.* **2017**, *147*, 371–379. [CrossRef]
43. Cheng, X.; Long, D.; Zhang, C.; Gao, X.; Yu, Y.; Mei, K.; Zhang, C.; Guo, X.; Chen, Z. Utilization of red mud, slag and waste drilling fluid for the synthesis of slag-red mud cementitious material. *J. Clean. Prod.* **2019**, *238*, 117902. [CrossRef]
44. Huang, Z.; Liu, K.; Duan, J.; Wang, Q. A review of waste-containing building materials: Characterization of the heavy metal. *Constr. Build. Mater.* **2021**, *309*, 125107. [CrossRef]
45. Lee, W.K.W.; van Deventer, J.S.J. The effect of ionic contaminants on the early-age properties of alkali-activated fly ash-based cements. *Cem. Concr. Res.* **2002**, *32*, 577–584. [CrossRef]
46. Yunsheng, Z.; Wei, S.; Qianli, C.; Lin, C. Synthesis and heavy metal immobilization behaviors of slag based geopolymer. *J. Hazard. Mater.* **2007**, *143*, 206–213. [CrossRef]

Article

Mechanical Properties and Microstructure of Alkali-Activated Soda Residue-Blast Furnace Slag Composite Binder

Zhaoyun Zhang [1,2,*], Chuang Xie [1,*], Zhaohu Sang [2] and Dejun Li [2]

1 School of Chemical Engineering and Technology, Tianjin University, Tianjin 300072, China
2 Tangshan Sanyou Alkali Chloride Co., Ltd., Tangshan 063000, China
* Correspondence: zhang924001@163.com (Z.Z.); acxie@tju.edu.cn (C.X.)

Abstract: This study prepared an alkali-activated soda residue (SR)-blast furnace slag (BFS) composite binder by adding a large amount of SR to the alkali-activated material system. Considering many factors, such as the Na_2O content, ratio of SR to BFS and the water-binder ratio, the variation patterns in the new binder's mechanical properties and its micro-evolution mechanisms were assessed. The results show that the compressive strength first grew and then dropped with the Na_2O content, with an optimal level at 3.0%. At this level, the strength values of the 3d and 28d samples were 10.5 and 27.8 MPa, respectively, exceeding those in the control group without Na_2O by 337.5 and 69.5%, respectively. As the Na_2O admixture increased from 0 to 3%, the fluidity of the mortar decreased from 156 mm to 127 mm due to the high frictional resistance caused by the faster generation of hydration products, and the high water absorption of SR also led to reduced fluidity. The new binder's hydration process mainly generated C-(A)-S-H gel, ettringite (ET), hydrocalumite (HC), calcium hydroxide (CH), and other crystalline hydrates. A 3% Na_2O content inhibited the ET growth but significantly promoted the formation of uniformly distributed C-(A)-S-H gel and HC. Crystals grew in the pores or were interspersed in the gel, filling microcracks and significantly increasing the structure density and strength. Excessive Na_2O (>3%) could promote the generation of non-uniformly distributed gel, producing more macropores in the matrix and reducing its strength. Additionally, the increased SR content was not conducive to C-(A)-S-H gel formation, but significantly promoted ET formation, which would inhibit strength development. This study provides a theoretical basis for replacing cement with this new binder in pavement bricks and other unreinforced products.

Keywords: composite binder; soda residue; blast furnace slag; mechanical properties; microstructure

Citation: Zhang, Z.; Xie, C.; Sang, Z.; Li, D. Mechanical Properties and Microstructure of Alkali-Activated Soda Residue-Blast Furnace Slag Composite Binder. *Sustainability* **2022**, *14*, 11751. https://doi.org/10.3390/su141811751

Academic Editor: Gianluca Mazzucco

Received: 12 August 2022
Accepted: 12 September 2022
Published: 19 September 2022

Publisher's Note: MDPI stays neutral with regard to jurisdictional claims in published maps and institutional affiliations.

Copyright: © 2022 by the authors. Licensee MDPI, Basel, Switzerland. This article is an open access article distributed under the terms and conditions of the Creative Commons Attribution (CC BY) license (https://creativecommons.org/licenses/by/4.0/).

1. Introduction

After decades of development, alkali-activated binders are considered the most lucrative cement substitute in construction. Compared with the complicated manufacturing process of cement, alkali-activated binders have the advantages of low energy consumption, fewer CO_2 emissions, and substantial environmental benefits [1,2]. Among various alkali-activated binders, the alkali-activated blast furnace slag binder (AAS) has been extensively studied worldwide. As early as 1940, Purdon [3] used different activators (caustic alkali such as NaOH, other alkaline salts, etc.) to activate various types of BFS, systematically studied the strength development of AAS, and found that it had better tensile and flexural properties and low hydration heat release. Subsequently, this kind of binder has become a strong focus for research.

The mechanical properties of AAS are related to many factors, such as the physical and chemical properties of raw materials (e.g., fineness, chemical composition, activity), the type and content of activator, and curing conditions, etc. [4,5]. Generally, AAS has a short setting time and rapid early strength development [6,7], and its one-day strength can reach more than 40 MPa, which is due to the rapid hydration of BFS attributable to high pH activation. Marvila et al. [7] investigated the feasibility of using alkali activated cements based on BFS as an alternative to ordinary Portland cement. The results showed that the system had the highest compressive strength at 10% Na_2O, and had lower porosity and better durability. Asaad et al. [8] prepared a new AAS using different levels (5, 10, 15, 20, 25%) of metakaolin instead of BFS and found that the 28-day compressive strength of the mortar samples with 10% metakaolin was 63.4 MPa. Metakaolin addition significantly reduced the drying shrinkage of the samples, decreased the porosity and carbonation depth and improved the corrosion resistance. In recent years, some solid waste materials have been introduced into AAS to prepare new composite binders, reducing the cost and enabling the reuse of other solid wastes. Islam et al. [9] used $NaOH + NaSiO_3$ to activate the composite precursor material of palm oil fuel ash and BFS followed by curing at 65 °C for 24 h. They reported that the optimal ratio of palm oil fuel ash and BFS was 3:7, and the 28d strength could reach 66 MPa. Nasir et al. [10] also used $NaOH + NaSiO_3$ to activate the composite precursor material containing 70% silicon manganese soot and 30% BFS. It was found that curing at 60 °C for 6 h was the optimal curing condition, and the compressive strength values at 3d and 28d were 37.8 and 45.2 MPa, respectively. The addition of other solid wastes could affect the mechanical properties and hydration process of AAS and address some disadvantages of AAS, such as delayed setting time. Therefore, this kind of new composite binder has gradually become a strong research focus.

Soda residue (SR) is the waste residue produced in the soda industry of the ammonia alkali process. Its main chemical components include CaO, a small amount of SiO_2, Al_2O_3, and a high chloride content [11,12]. Each ton of soda produced in this process generates 300–600 kg of SR, and over 7 million tons of solid SR are generated annually in China [13]. Due to this large volume of SR and the few ways the resource can be utilized, most factories mainly deal with SR by outward transportation and accumulation, which requires much land and results in soil and groundwater pollution [14]. Some factories even directly discharge the SR into the sea, polluting the seawater and jeopardizing its flora and fauna. Therefore, more effective and safe utilization of SR is very topical from both environmental and economic standpoints.

Scholars have recently conducted resource utilization research on SR based on its physical and chemical properties, mainly in the areas of soil improvement, harmful ion adsorption, and building material manufacturing. In terms of soil improvement, the main chemical composition of SR is insoluble salts. After mixing with fly ash, cement and other materials, it can form SR soil, which can be used for low-lying land filling, underground tunnel backfilling or road foundations [15,16]. Since the SR's solid structure comprises aggregates with many pores and good adsorption properties, it can be used for desulfurization or adsorption of harmful ions in various metallurgical industries [17]. In addition, SR is rich in Ca, Si, and Al, making it similar to cement clinker and applicable to cement production. Thus, Uçal et al. [18] prepared Alinite cement with SR as a raw material and reported that the prepared cement had a shorter induction period (between 15 and 20 min) than ordinary Portland cement. The addition of 12% gypsum could increase the amount of ettringite in hydration products and significantly improve the compressive strength. Wang et al. [14] mixed SR with cement, ground it and calcined it to prepare composite cement. At a 5% SR content, the 28d compressive strength reached 53 MPa and the contents of CaO and Cl^- in the clinker met the requirements of relevant standards, but the frost resistance and impermeability were weakened.

In recent years, the research on the preparation of new binders with SR has gradually increased. It mainly includes the following two main approaches: (1) using SR as an activator, since the alkali contained in SR could replace strong alkali to activate aluminosilicate raw materials, and (2) preparation of composite binders by adding SR to conventional alkali-activated materials. Within the framework of the first approach, Liu [19] and Lin [20] adopted SR to activate BFS and prepare binders with a 28d strength exceeding 30 MPa. However, their 3d and 7d strength values were less than 5 MPa, which limited their engineering application. Besides, the contents of SR were small (less than 20%), which implied insufficient SR utilization. Alternatively, Guo et al. [21,22] used SR and carbide slag to activate BFS-fly ash for preparing a binder system with strengths ranging from 17 to 40 MPa, but this preparation also only required a low content of SR. In the second approach, which involves adding SR to conventional alkali-activated materials, Zhao et al. [16] added SR to a sodium silicate-activated fly ash cementitious system. At an SR content of 40%, a 2 mol/L concentration of sodium silicate solution, and the liquid-solid ratio of 1.2, relatively good fluidity and compressive strength (3.7 MPa) were obtained, which were sufficient for goaf backfilling applications. As for the hydration products of SR-based binders, high chloride contents have been reported to promote the formation of Friedel's salt, improving the matrix strength [21,23,24].

The above brief survey strongly indicates that the SR-based preparation of clinker-free binders is still in the preliminary exploration stage. In most studies, the amounts of SR were too small to mitigate the environmental problems caused by the rapid accumulation of SR. Therefore, in this study, a large amount of SR was introduced into the alkali-activated material system to prepare an alkali-activated SR-BFS composite binder. The optimum mechanical properties were obtained by optimizing the activator content, and the microstructure evolution process was systematically studied to reveal the hydration mechanism. This study aims to provide a solution to the problems of environmental pollution and land waste caused by the large-scale accumulation of SR.

2. Materials and Methods

2.1. Raw Materials

The raw materials included SR, BFS, NaOH, standard sand, and water. The chemical composition, particle size distribution, micromorphology, and phase composition of SR and BFS were assessed via XRF, laser particle size analyzer, SEM, and XRD, respectively. The specific surface area was measured by gas adsorption using the BET method, and the density was assessed via the density bottle method. The results are summarized in Table 1 and plotted in Figures 1–3.

Table 1. Chemical composition and physical properties of raw materials.

Chemical Composition (wt/%)	SR	GGBS
Ca	43.2	33.7
Si	9.87	32.6
Al	3.25	17.1
Fe	0.91	1.18
Mg	9.77	7.40
Ti	0.12	1.54
K	0.29	0.57
S	5.57	3.21
Na	3.93	0.56
Cl	20.23	-
Loss on ignition	2.86	2.14
Physical properties		
Specific gravity (kg/m^3)	2351	2742
Surface area (m^2/kg)	261.2	419.5

The SR with an initial moisture content exceeding 90% was acquired from the Tangshan Sanyou Alkali-Chlor Co., Ltd. (China). The SR samples were prepared as follows. First, the samples were left to stand for layering to occur. The clear upper liquid was then removed and the sedimentary layer was stirred evenly. When the moisture content of the sedimentary layer was reduced to 50–60%, the sediments were oven-dried, ground, and sieved with a 0.15 mm square hole sieve to obtain SR powder with an average particle size of 20.31 μm (Figure 1). The density of the SR powder was 2351 kg/m^3, and the specific surface area was 261.2 m^2/kg (Table 1). The micromorphology was agglomerated with many pores, as shown in Figure 2a. The main chemical components were CaO and Cl$^-$, and small amounts of SiO_2, Al_2O_3, MgO (Table 1). The main phases were calcium chloride hydroxide, gypsum, halite, and calcite (Figure 3a). The S95 grade BFS was utilized and had an average particle size of 14.60 μm (Figure 1), a 2742 kg/m³ density, and a 419.5 m²/kg specific surface area. The main chemical components were CaO, SiO_2, and Al_2O_3 (Table 1), and the main phase was quartz (Figure 3b). As seen in Figure 2b, the micromorphology was irregular blocks. NaOH was a granular analytical reagent with a content exceeding 96%. The acquired standard sand, with a particle size range of 0.08–2.0 mm and mud content below 0.2%, complied with the Chinese standard GB/T17671. Tap water from the laboratory was used to ensure the required water-binder ratios.

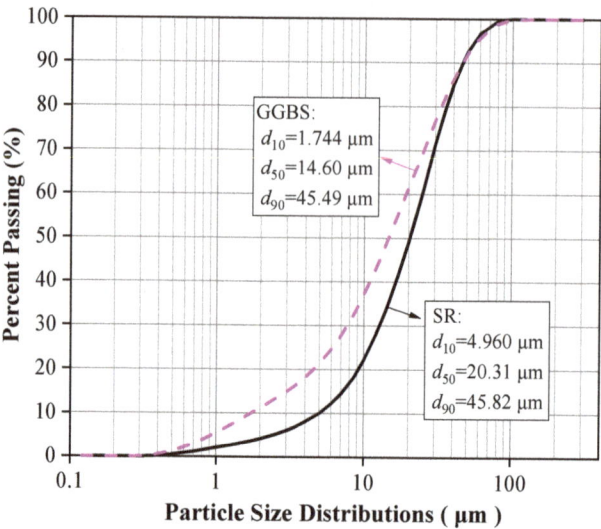

Figure 1. Particle size distributions of SR and GGBS.

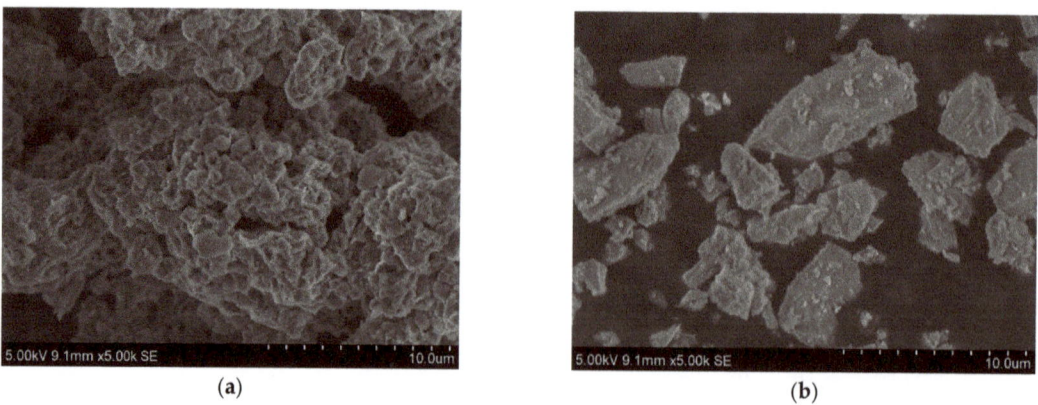

Figure 2. SEM images of (a) SR, (b) GGBS.

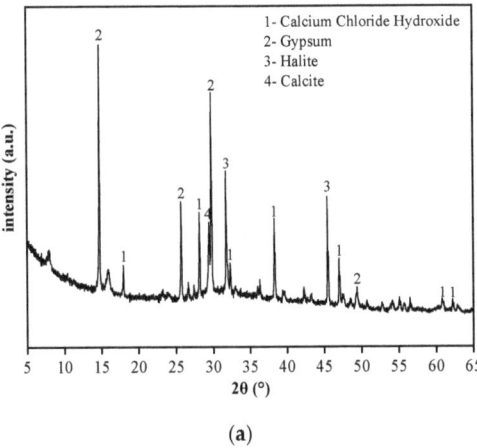
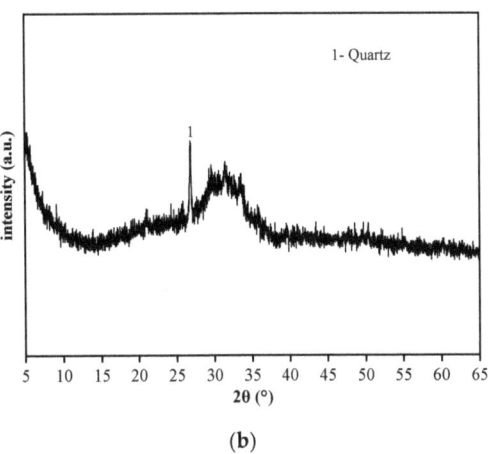

Figure 3. XRD spectra of (a) SR, and (b) GGBS.

2.2. Mixing Ratio Design

This study introduced a large amount of SR into AAS to prepare a new composite binder. Through preliminary exploratory tests, the initial SR and BFS shares were set to 40 and 60%, respectively, to ensure the maximal long-term strength. Firstly, the Na_2O content effect on the binder's properties was investigated. The Na_2O content was defined as the ratio of Na_2O mass contained in NaOH to the mass of binders. The designed contents of Na_2O were 0, 0.5, 1.0, 1.5, 2.0, 3.0 and 4.0%. The optimum content of Na_2O was determined according to the maximal compressive strength. On this basis, the effects of the ratio of SR to BFS and water-binder ratio on the binder's properties were further studied. In compliance with the evaluation method of Portland cement performance in Chinese standard GB/T 17671 [25], the water-binder ratio was fixed at 1:2, and the binder-sand ratio was fixed at 1:3. The mix proportions of mortar and paste are shown in Table 2. The mortar samples were tested to assess the compressive strength and fluidity, and the paste samples were used for XRD, TG-DTG, FTIR, and SEM analyses.

Table 2. Mix proportion design.

Group		Content of Raw Materials in Binder/g			Na_2O Content (%)	Standard Sand /g	Water-Binder Ratio
		SR	GGBS	NaOH			
Mortar	SG	180	270	0	0	1350	0.50
	SGN1	180	270	2.9	0.5	1350	0.50
	SGN2	180	270	5.8	1.0	1350	0.50
	SGN3	180	270	8.7	1.5	1350	0.50
	SGN4	180	270	11.6	2.0	1350	0.50
	SGN5	180	270	17.4	3.0	1350	0.50
	SGN6	180	270	23.2	4.0	1350	0.50
	SGN7	225	225	17.4	3.0	1350	0.50
	SGN8	270	180	17.4	3.0	1350	0.50
	SGN9	180	270	17.4	3.0	1350	0.40
	SGN10	180	270	17.4	3.0	1350	0.45
	SGN11	180	270	17.4	3.0	1350	0.55
Paste	SG	180	270	0	0	0	0.50
	SGN2	180	270	5.8	1.0	0	0.50
	SGN5	180	270	17.4	3.0	0	0.50
	SGN6	180	270	23.2	4.0	0	0.50
	SGN8	270	180	17.4	3.0	0	0.50

2.3. Preparation of Samples

The preparation process of the paste and mortar samples was as follows. The amount of water required for the particular water-binder ratio was calculated before the test. Then, the NaOH particles were dissolved in the water and left for 2–4 h. During the test, the NaOH solution was first poured into the mixing pot, then, the SR and BFS were added. The mixture was stirred at a 140 r/min rotation rate for 1 min, and then was stirred at 285 r/min for 2 min to obtain the paste. At this point, the standard sand was added to the mixture. The mixture was stirred at 140 r/min for 30 s, and then was stirred at 285 r/min for 1.5 min to obtain the mortar.

After the mixing process was completed, the paste (or mortar) was loaded into the test mold in two layers, and the mold was vibrated for 60 s on the vibration table after each layer was loaded. The surface was smoothed with a leveling ruler. After the pouring process, the test mold was placed into a curing box with a temperature of $20 \pm 1\,°C$ and relative humidity of no less than 90% for 24 h. Then, the test mold was removed, and the samples were numbered and cured in the curing box until the predetermined age.

2.4. Experimental Method

In this study, the macro-performance tests measured the mortar samples' compressive strength and fluidity. The microstructural tests including XRD, TG-DTG, FTIR, and SEM were performed on the paste samples. The experimental flowchart is shown in Figure 4.

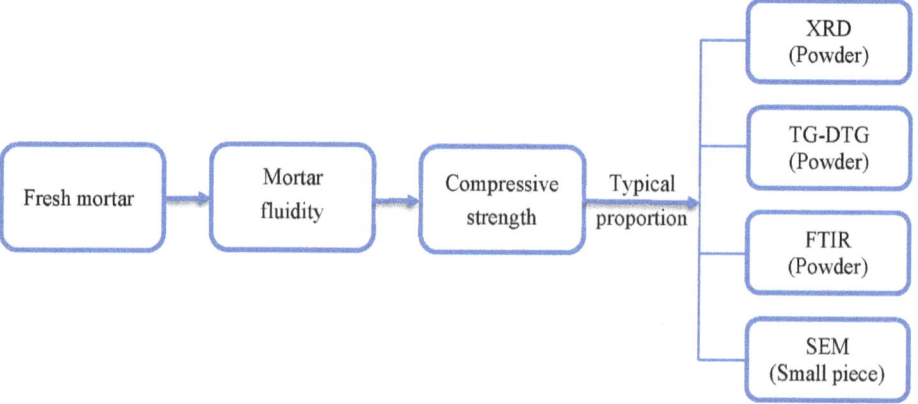

Figure 4. Experimental flowchart.

According to Chinese standard GB/T 17671 [25], the prism-shaped mortar samples with 40 mm × 40 mm × 160 mm dimensions should be used for the compressive strength test. The curing ages were 3, 7, 28, 56d, and three specimens were selected for each mixing ratio and curing age. The samples were subjected to static compressive loading at a loading rate of 2.4 ± 0.2 kN until they failed. The average and standard deviations of six test values on three test samples were taken as the test results. The mortar fluidity tests complied with the Chinese standard GB/T 2419 [26]. The fresh mortar was loaded into the truncated cone circular mold in two layers, the first and second layers being uniformly tamped fifteen and ten times, respectively. Then, the mold was lifted; after 25 jumps on the jumping table, the bottom diameter of the mortar was measured with a ruler. Each mixing ratio was tested three times, and the average value and standard deviation were taken as the test results.

Cubic paste samples with 50 mm × 50 mm × 50 mm dimensions were prepared for the microscopic tests. The test samples were broken into blocks at the predetermined curing age and dried by the following procedure [27]. They were first soaked in isopropanol solution for 24 h and then soaked for 6d after replacing the isopropanol solution. Finally, the samples were vacuumed continuously in a vacuum dryer until the solvent volatilized fully. For the XRD, TG-DTG, FTIR tests, the dry blocks were ground into powder and passed through a 0.075 mm square-hole sieve. The maximum side length of the block in the SEM test did not exceed 10 mm. The XRD test parameters were as follows: Cu target, tube voltage of 40 kV, tube current of 40 mA, scanning angle range of 5~65°, and scanning speed of 2°/min. The TG-DTG parameters were as follows: a temperature range from room temperature to 1000 °C, Argon as a shielding gas, and a heating rate of 10 °C/min. The FTIR test involved a wavenumber range of 400 ~ 4000 cm^{-1} and a 2 cm^{-1} resolution. The SEM test was conducted via a TESCAN-VEGA3 scanning electron microscope.

3. Results and Discussion

3.1. Compressive Strength

Figure 5 shows the compressive strength values of all the tested mortar samples of various curing ages. The effects of Na$_2$O content, SR to BFS ratio, and water-binder ratio on the compressive strength were considered.

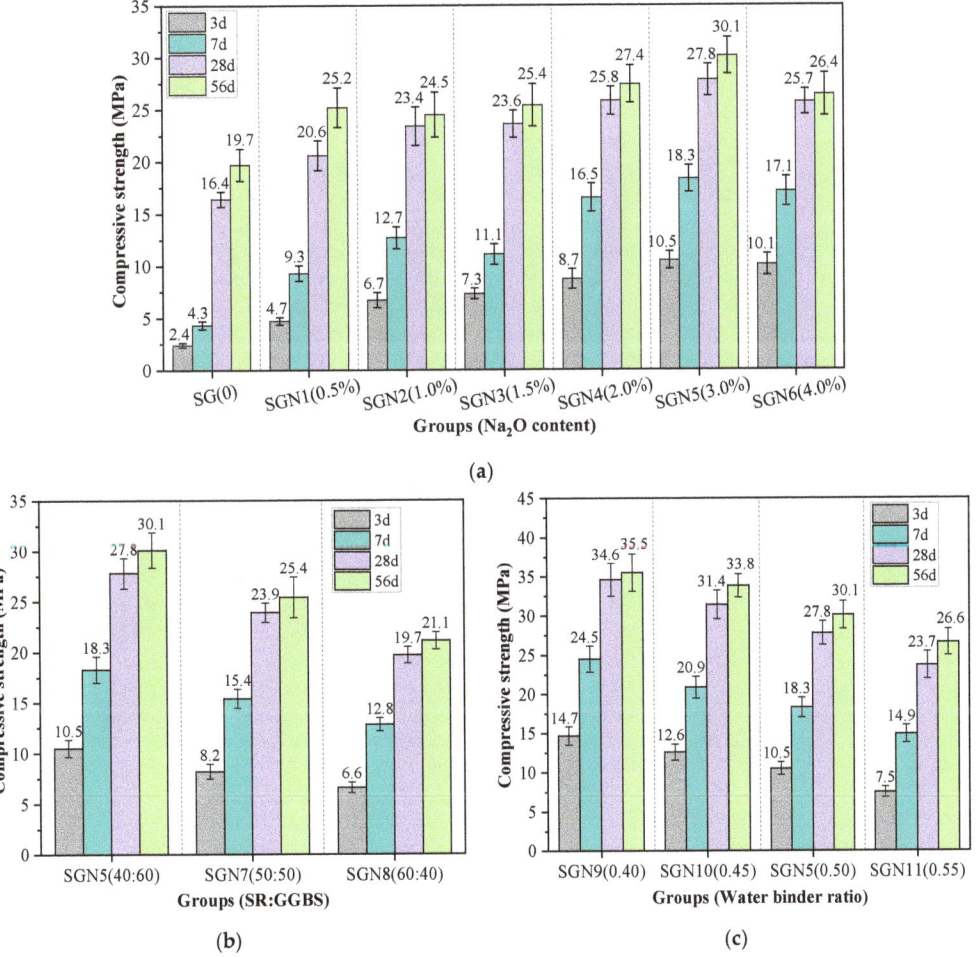

Figure 5. Test results of compressive strength: (**a**) effect of Na$_2$O content; (**b**) effect of SR to GGBS ratio; and (**c**) effect of water-binder ratio.

According to Figure 5a, the early strength development of SR-BFS composite binder (SG) without NaOH was very slow, and the 3d and 7d strength values were only 2.4 and 4.3 MPa, respectively. However, with increased curing age, the 28d and 56d strength reached 16.4 and 19.7 MPa, respectively, indicating that the alkalinity in SR stimulated the activity component of the SR-BFS composite precursor. Still, this excitation effect was mainly observed in the later stage, similar to previous research results [19,20]. When the NaOH (Na_2O) was mixed with SG, the compressive strength at each curing age grew first and then dropped with increased amounts added, similar to previous research results [7]. At an Na_2O content of 3.0%, the early and late strengths were the highest, the 3d and 28d strengths were 10.5 and 27.8 MPa, respectively, exceeding those of SG mixtures by 337.5% and 69.5%, respectively. The possible reason was that the addition of NaOH significantly enhances the system's alkalinity, promoting the activation of the active components of composite precursor and accelerating the formation of C-(A)-S-H gel [28]. This rapidly improves the strength. Notably, excessive NaOH could reduce the strength, and this finding was consistent with previous studies [29,30]. The reason may be that although too high alkalinity can promote rapid hydration of the precursors, the excessive formation rate of gel products leads to their inadequate dissolution, transformation, and precipitation, resulting in rapid accumulation and encapsulation in the precursor particles. The gel is then unevenly distributed, forming more pores and preventing further strength development [30,31].

According to Figure 5b, with an increase in the SR-BFS ratio, the compressive strength at each curing age decreased gradually. This is because BFS is the main precursor in SR-BFS composite precursors, and the BFS decline means the decrease of active substances; the amount of gel products in the hydration reduces, resulting in strength reduction. Although the $CaSO_4$ in the SR can promote the formation of ettringite in the system (conducive to early strength growth) [32], and the $CaCO_3$ crystals in the SR also have a microaggregate effect (improving the structural compactness) [33], these effects did not play a decisive role in the strength development. According to Figure 5c, the compressive strength at each curing age decreased gradually with an increase in the water-binder ratio, which is consistent with that of cement-based materials. This is because the increase in the water-binder ratio increases the free water content in the system. Thus, the number of blisters formed by excess water after hardening or pores formed by water evaporation increases, reducing the structural compactness and causing a decrease in strength.

3.2. Mortar Fluidity

Mortar fluidity reflects the water demand of new binders and is also an important workability index of binders. Figure 6 shows the fluidity test results for all the tested fresh mortar samples.

It can be seen in Figure 6 that the fluidity of the SG group without NaOH was the highest, namely 159 mm. The fluidity decreased with the increase in NaOH content. At a Na_2O content of 3.0%, the fluidity decreased to 127 mm, i.e., by 20.1%. This may be caused by the following reasons: the NaOH accelerates the hydration reaction of the SR-BFS composite precursors so that the hydration products (such as C-(A)-S-H gel and others) can be rapidly generated, increasing the internal friction resistance of the mortar and reducing the fluidity [34,35]. According to the results on SGN5, SGN7 and SGN8 samples, with an increase in the SR-BFS ratio, the fluidity of the mortar gradually decreased from 127 to 115 and then to 104 mm. This is because the SR particles are prone to agglomerate, rich in pores, and the water absorption ability is strong. When the SR content increases, the number of adsorbed water molecules increases under the same water consumption, reducing the free water content and fluidity. According to the SGN5 and SGN9–11 results, the fluidity grew with the water-binder ratio, which is consistent with the properties of cement. This is because the increased water-binder ratio significantly increases the amount of free water in the mortar, weakens the internal friction of the mortar, and improves the fluidity.

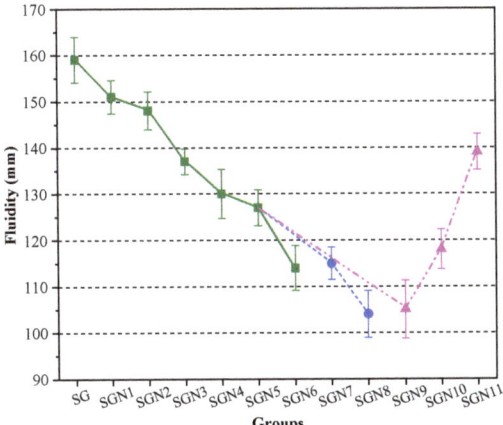

Figure 6. Test results of mortar fluidity.

3.3. XRD Analysis

To determine the hydration products of the new binders and characterize the effects of Na$_2$O content and SR-BFS ratio on the hydration products, XRD tests at the age of 28d were conducted on five groups of samples (SG, SGN2, 5, 6, 8). The test results are shown in Figure 7.

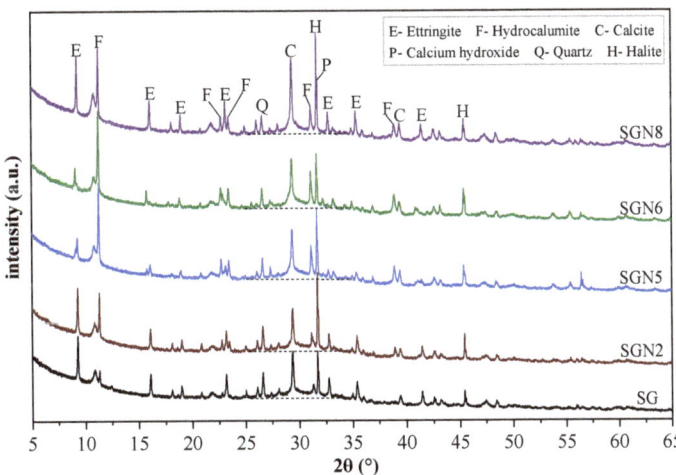

Figure 7. XRD patterns of 28d samples.

As seen in Figure 7, six main crystal phases were detected, among which ettringite (3CaO·Al$_2$O$_3$·3CaSO$_4$·32H$_2$O), hydrocalumite (3CaO·Al$_2$O$_3$·CaCl$_2$·10H$_2$O, Friedel's salt), and calcium hydroxide [Ca(OH)$_2$] were the main crystal hydration products. These were generated by Cl$^-$, CaSO$_4$, CaO, and Al$_2$O$_3$ provided by SR-BFS composite precursors under the activation of NaOH. Quartz (SiO$_2$) emerged from the BFS, and calcite (CaCO$_3$) from the SR or carbonation of Ca(OH)$_2$. Halite (NaCl) was formed by combining free Cl$^-$ and Na$^+$ ions.

According to Figure 7, the main hydration products, namely ettringite (ET), hydrocalumite (HC), and calcium hydroxide (CH), showed obvious variation. In the SG group (without NaOH), the diffraction peak intensity of ET was strong, in contrast to HC and CH.

When 1.0% Na$_2$O was added to SG, the peak intensity of ET decreased slightly, whereas those of HC and CH increased significantly. When the content of Na$_2$O was increased to 3.0%, the peak intensity of ET decreased significantly, whereas that of HC increased significantly, and that of CH changed little. This indicates that the increased amount of NaOH inhibited the ET growth but promoted the growth of HC and CH. When the content of Na$_2$O was further increased to 4.0%, the peak intensities of ET, HC, and CH no longer changed significantly, indicating that excessive Na$_2$O contents no longer played a role in the variation of hydration products. By comparing SGN5 and SGN8 samples, the peak intensities of ET and CH in SGN8 increased significantly. Still, the peak intensity of HC did not change significantly, indicating that an increase in the SR content promoted the formation of ET and CH but had little effect on HC. Previous studies [29,36] have shown that the early growth of ET can promote hardening, which is conducive to early strength formation, but the later formed high-content ET tends to expand and cause cracking, harming the late strength formation. The growth of HC can better fill the matrix pores, which is generally considered beneficial to the strength development [23]. Combined with the strength development described in Section 3.1, it can be seen that a suitable amount of NaOH can promote the growth of the favorable crystal product HC and inhibit the growth of the harmful crystal product ET, which makes a certain contribution to the strength development.

The diffraction peak intensity of quartz decreased gradually with Na$_2$O content because the high alkalinity of NaOH significantly promoted the dissolution of the quartz phase, in which silicon participated in the formation of C-(A)-S-H gel [31,37]. Especially when the Na$_2$O content increased from 1.0 to 3.0%, the peak intensity decreased significantly, indicating that the gel production increased significantly. However, the peak intensity remained unchanged when the content reached 4%, indicating that the promotion effect on gel formation was saturated. The quartz peak intensity of SGN8 was the lowest due to the low content of BFS in the system. With an increase in Na$_2$O content, the diffraction peak intensity of calcite first decreased slightly and then increased gradually. The reason might be that when a small amount of NaOH was introduced into the system, a small amount of calcite was decomposed into Ca^{2+}, participating in the hydration reaction. When more NaOH was added, it was prone to carbonation during solidification and sample treatment, increasing the amount of calcite. The calcite peak of SGN8 was the strongest, obviously caused by the high content of SR in the system. The diffraction peak intensity of halite increased first and then decreased with Na$_2$O content. At an Na$_2$O content of 1.0%, the peak intensity of halite increased significantly compared with SG because NaOH introduced a large amount of Na$^+$, significantly promoting the formation of halite. When the content of Na$_2$O was increased to 3.0%, and then to 4.0%, the peak intensity of halite gradually decreased, which might be because the high content of NaOH significantly promoted the formation of HC and consumed more Cl$^-$, thus inhibiting the formation of halite. The halite peak of SGN8 was the strongest, obviously due to the high content of SR and NaOH in the system, and more halite was formed by Cl$^-$ and Na$^+$ ions.

Additionally, "humps" representing the amorphous phase appeared in the 25–35° range of all the diagrams, indicating that the amorphous C-(A)-S-H was formed by a hydration reaction [29,38,39]. With increased Na$_2$O content, the width and height of these "humps" increased, indicating that the gel production was enhanced, which is conducive to a growth in strength.

3.4. TG-DTG Analysis

To further determine the composition and evolution patterns of hydration products in binders, TG tests at 28d curing age were conducted for five groups of samples (SG, SGN2, 5, 6, 8). Figure 8 shows the TG-DTG curves obtained in the tests. Figure 9 shows the statistical weight loss results in different temperature ranges.

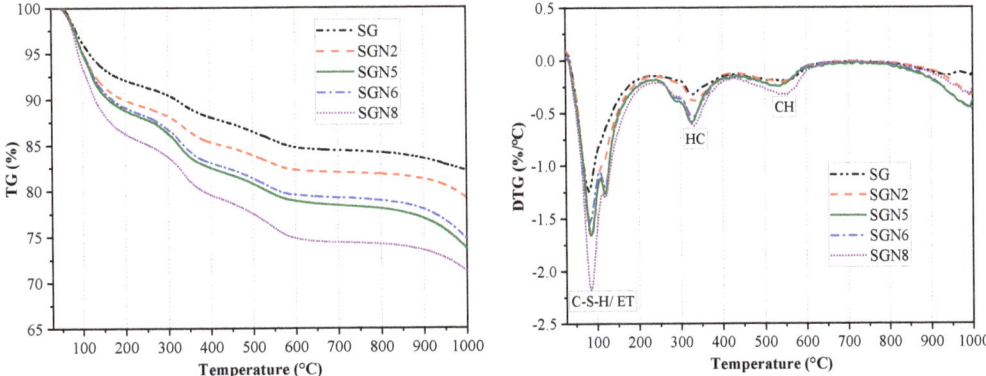

Figure 8. TG–DTG curves of 28-day samples.

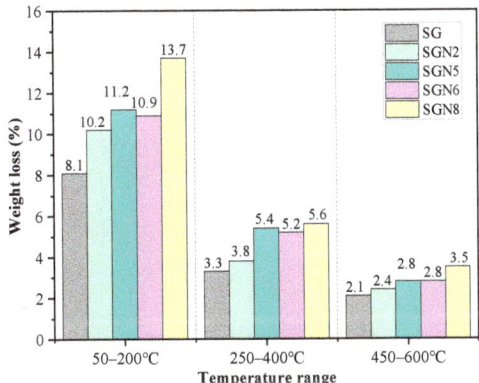

Figure 9. Weight loss at different temperature ranges for specimens at 28 days.

According to Figure 8, there were mainly three weight loss peaks. The weight loss at 50–200 °C was caused by the dehydration of C-(A)-S-H gel and ET [40,41]. According to a comprehensive analysis of the test results depicted in Figure 9 and XRD examinations, it was revealed that when a 1.0% content of Na_2O was added to SG, the intensity of the weight loss peak significantly increased with the weight loss increasing from 8.1 to 10.2%, indicating that the amount of C-(A)-S-H gel produced by hydration was significantly increased. When the content of Na_2O was increased to 3.0%, the weight loss further grew to 11.2%. As the amount of ET in the hydration products decreased significantly (Figure 7), the increase of this weight loss was still caused by the significant increase of C-(A)-S-H gel produced by hydration. When the Na_2O content was further increased to 4.0%, the weight loss was unchanged, indicating that the excessive alkali content had no obvious effect on C-(A)-S-H gel and ET formation. The intensity of this weight loss peak of SGN8 was the strongest, and the weight loss (13.7%) significantly exceeded that of SGN5, which was mainly caused by the large increase in ET production.

The dehydration and decomposition of HC caused the weight loss at 250–400 °C [14,33]. When 1.0% and 3.0% Na_2O were added to SG, the intensity of the weight loss peak increased gradually. The weight loss grew from 3.3% to 3.8% and 5.4%, indicating that the amount of HC in the hydration products increased gradually. When the content of Na_2O was increased to 4.0%, the weight loss was unchanged, indicating that excessive alkali contents had no obvious effect on the formation of HC. There was no obvious change in the weight loss of SGN8 compared with SGN5, indicating that the increase of SR content had no obvious effect on the formation of HC crystals. The variation pattern of HC crystals in the TG-DTG test was consistent with that observed via XRD.

The weight loss at 450–600 °C was caused by CH decomposition [29,40]. With the addition of 1.0 and 3.0% Na$_2$O to SG, the weight loss gradually increased from 2.1 to 2.4 and 2.8%, respectively, indicating that the CH crystals in the hydration product gradually increased. However, increasing Na$_2$O content to 4.0% had no obvious effect on the formation of CH because the weight loss remained unchanged. The intensity of this weight loss peak of SGN8 was the strongest, and the weight loss (3.5%) was the largest, indicating that the most CH was generated in the hydration reaction. The variation pattern of CH in the TG-DTG test was consistent with that in XRD.

3.5. FTIR Analysis

Fourier transform infrared spectroscopy (FTIR) provides a useful tool for qualitative and quantitative analysis of hydration products of binders. FTIR tests of 28d samples were conducted on five groups of samples (SG, SGN2, 5, 6, and 8). The test results are plotted in Figure 10.

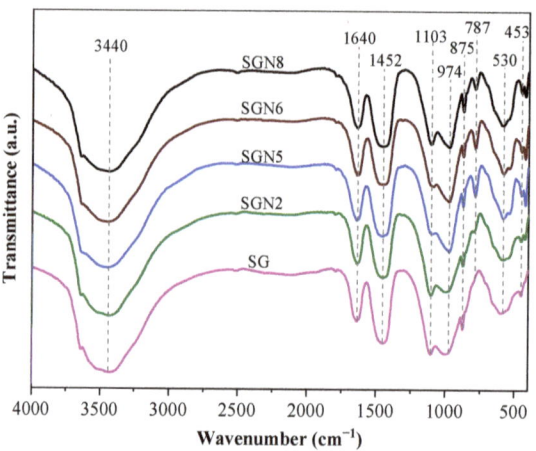

Figure 10. FTIR curves of 28 d samples.

Ten main absorption peaks can be observed in Figure 10. The absorption peak near 452 cm^{-1} was caused by the in-plane bending vibration of the Si-O bond. The Si-O bond's stretching vibration caused the absorption peaks near 797 and 974 cm^{-1}, and the Si-O-Al bond's deformation vibration caused the absorption peak near 583 cm^{-1}. These four absorption peaks indicate the existence of C-(A)-S-H gel in the hydration products [42,43]. The intensities of these absorption peaks were stronger in SGN5 and SGN6, slightly lower in SGN8, and the lowest in SG and SGN2. The results show that the increased Na$_2$O contents significantly promoted the hydration reaction and C-(A)-S-H gel generation, but the excessive Na$_2$O contents had no obvious promotion effect. Additionally, the increase of SR content was not conducive to C-(A)-S-H gel formation.

The absorption peak near 1103 cm^{-1} was caused by the stretching vibration of SO$_4^{2-}$ [44], indicating the existence of ET in the hydration products. The absorption peak intensity was the weakest in SGN5 and SGN6, stronger in SG, SGN2, and the strongest in SGN8, indicating that increased Na$_2$O content did not promote the formation of ET, in contrast to increased SR contents. The out-of-plane bending vibration of CO$_3^{2-}$ and the C-O bond's stretching vibration caused the absorption peaks near 875 and 1452 cm^{-1} [42,45], respectively, indicating the presence of calcite in these samples [46]. The intensities of these absorption peaks (and thus, the amount of calcite) decreased first and then increased with the Na$_2$O content, which is consistent with the XRD results. Additionally, the higher content of SR in SGN8 led to the highest calcite content and the highest corresponding absorption peak intensity in this sample. The absorption peaks near 1640 cm^{-1} and 3440 cm^{-1} were caused by the H-O-H bond's bending and stretching vibrations, which were the crystal water's internal vibration absorption characteristics [33,47]. The crystal water mainly came from C-(A)–S-H gel, HC, and ET [33]. The intensities of these two absorption peaks exhibited no obvious patterns related to the large difference in the production of the three hydration products in different groups. The absorption peak near 3641 cm^{-1} was caused by the vibration of the O-H bond in Ca(OH)$_2$ [48], indicating that there was CH in the hydration products, but the intensity change was not significant. The above results further proved the experimental results of XRD and TG-DTG.

3.6. SEM Analysis

To further determine the hydration products and the micromorphology evolution of the new binders, SEM and EDS tests of SG, SGN5, and SGN6 at the curing age of 28d were conducted. The results are depicted in Figure 11.

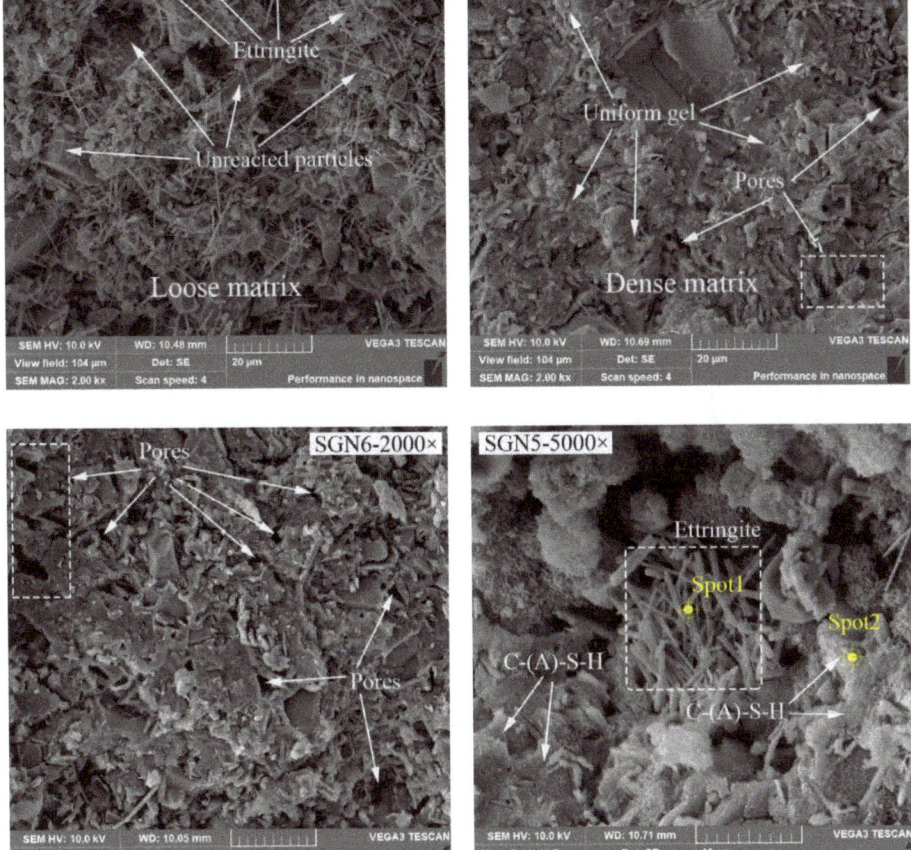

Figure 11. Cont.

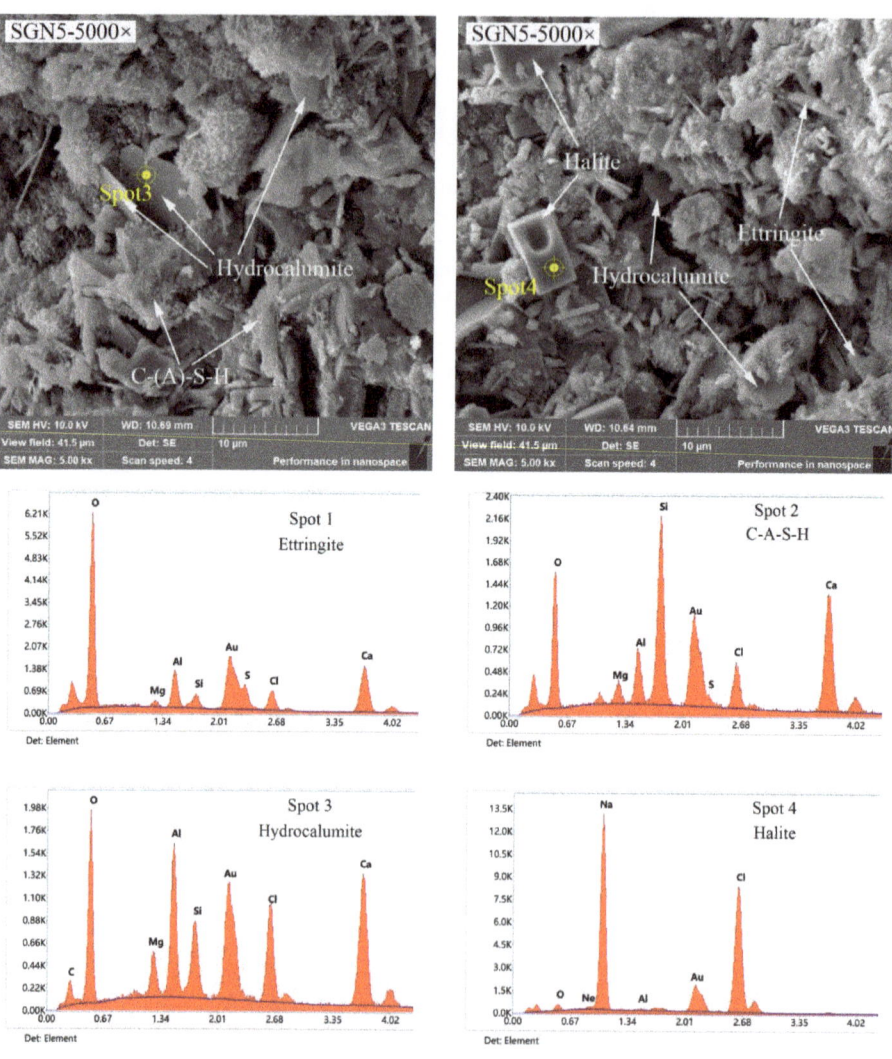

Figure 11. SEM and EDS images of 28d samples.

According to the SEM images of SG, SGN5, and SGN6 samples at a 2000 magnification, the SG's microstructure was loose, and the continuously distributed gel was scarce. In addition, there were many unreacted raw material particles, and a large amount of ET was generated in the system, which was unfavorable for the later strength development. When 3.0% Na_2O was added, the microstructure of SGN5 was significantly improved, more continuously distributed gels could be observed, the number of unreacted raw material particles was significantly reduced, and fewer macropores appeared. Thus, an appropriate amount of NaOH significantly promoted the hydration reaction and significantly improved the compactness. When 4% Na_2O was added, the number of macropores in the matrix increased obviously. This might be due to excessive alkalinity resulting in uneven distribution of rapidly generated gel, forming more macropores, which was an important reason for its late strength below SGN5.

The SEM images of the SGN5 sample at a 5000× magnification feature C-(A)-S-H gel, ettringite, hydrocalumite, halite, and other crystals. The ettringite was mainly in a needle-rod form [29,49], the hydrocalumite was mainly in a hexagonal flake form and appeared as irregular flakes when not completely grown [50]. The halite mainly presented as cuboids or cubes. These crystals grew in pores or were interspersed in gels, filling tiny cracks and increasing the structure compactness.

4. Conclusions

In this study, a new alkali-activated SR-BFS composite binder was prepared, and its mechanical properties and microstructure were systematically studied. It can provide solutions to the problems of environmental pollution and land waste caused by the large-scale accumulation of alkali residues. The following conclusions were drawn:

(1) At the SR to BFS ratio of 40:60, the compressive strength increased first and then dropped with the Na_2O content. At the optimum Na_2O content of 3.0%, the 3d and 28d strength values of 10.5 and 27.8 MPa exceeded those of the control group (without Na_2O) by 337.5 and 69.5%, respectively. The compressive strength decreased with increasing SR to BFS ratio and water-binder ratio.

(2) The increased Na_2O content reduced the mortar fluidity due to high friction caused by the rapid formation of hydration products promoted by high alkalinity. The high SR contents also reduced the mortar fluidity due to the high water absorption of SR.

(3) The C-(A)-S-H gel, ET, HC, CH, and other crystals were mainly formed in the hydration of the new binder. Increased Na_2O contents inhibited the formation of ET but significantly promoted the formation of C-(A)-S-H gel and HC, which was conducive to growth in strength. Excessive alkali contents had no obvious effect on the hydration products. The increased SR content was not conducive to the formation of C-(A)-S-H gel but promoted the formation of ET, which was unfavorable for strength development.

(4) Under the optimal Na_2O content, the gel distribution was more uniform, and crystals grew in the pores or interspersed in the gel, filling microcracks and increasing the structure compactness. Excessive Na_2O contents could speed up the hydration reaction, but the uneven distribution of rapidly generated gel induced more macropores in the matrix, reducing its strength.

Author Contributions: Data curation, Z.Z. and Z.S.; Formal analysis, Z.S.; Investigation, Z.Z. and D.L.; Methodology, Z.Z.; Project administration, C.X.; Resources, Z.S.; Validation, Z.Z., Z.S. and D.L.; Writing—original draft, Z.Z.; writing—review & editing, C.X. All authors have read and agreed to the published version of the manuscript.

Funding: This research was funded by the National Natural Science Foundation of China under the grant of 21776204.

Acknowledgments: Appreciated the Tangshan Sanyou Alkali Chloride Co., Ltd., for technical consulting, and Tianjin University for academic assist.

Conflicts of Interest: The authors declare no conflict of interest.

References

1. Provis, J.L.; Palomo, A.; Shi, C. Advances in understanding alkali-activated materials. *Cem. Concr. Res.* **2015**, *78*, 110–125. [CrossRef]
2. Luukkonen, T.; Abdollahnejad, Z.; Yliniemi, J.; Kinnunen, P.; Illikainen, M. One-part alkali-activated materials: A review. *Cem. Concr. Res.* **2018**, *103*, 21–34. [CrossRef]
3. Purdon, A.O. The action of alkalis on blast-furnace slag. *J. Soc. Chem. Ind.* **1940**, *59*, 191–202.
4. Pacheco-Torgal, F.; Castro-Gomes, J.; Jalali, S. Alkali-activated binders: A review—Part 1. Historical background, terminology, reaction mechanisms and hydration products. *Constr. Build. Mater.* **2008**, *22*, 1305–1314. [CrossRef]
5. de Azevedo, A.R.G. Special edition: Practical applications of durability of activated alkali and geopolymer materials. *Case Stud. Constr. Mater.* **2022**, *17*, e01161. [CrossRef]
6. Zheng, W.; Chen, W.; Wang, Y. High-temperature resistance performance of alkali-activated slag cementitious materials. *J. Huazhong Univ. Sci. Technol.* **2009**, *10*, 96–99.

7. Marvila, M.T.; Azevedo, A.R.G.; Oliveira, L.B.; Xavier, G.C.; Vieira, C.M.F. Mechanical, physical and durability properties of activated alkali cement based on blast furnace slag as a function of Na_2O. *Case Stud. Constr. Mater.* **2021**, *15*, e00723. [CrossRef]
8. Asaad, M.A.; Huseien, G.F.; Memon, R.P.; Ghoshal, S.K.; Mohammadhosseini, H.; Alyousef, R. Enduring performance of alkali-activated mortars with metakaolin as granulated blast furnace slag replacement. *Case Stud. Constr. Mater.* **2022**, *16*, e00845. [CrossRef]
9. Islam, A.; Alengaram, U.J.; Jumaat, M.Z.; Bashar, I.I. The development of compressive strength of ground granulated blast furnace slag-palm oil fuel ash-fly ash based geopolymer mortar. *Mater. Des.* **2014**, *56*, 833–841. [CrossRef]
10. Nasir, M.; Johari, M.A.M.; Maslehuddin, M.; Yusuf, M.O.; Al-Harthi, M.A. Influence of heat curing period and temperature on the strength of silico-manganese fume-blast furnace slag-based alkali-activated mortar. *Constr. Build. Mater.* **2020**, *251*, 118961. [CrossRef]
11. Zhao, X.H.; Liu, C.Y.; Wang, L.; Zuo, L.M.; Zhu, Q.; Ma, W. Physical and mechanical properties and micro characteristics of fly ash-based geopolymers incorporating soda residue. *Cem. Concr. Compos.* **2019**, *98*, 125–136. [CrossRef]
12. Song, R.; Zhao, Q.; Zhang, J.; Liu, J. Microstructure and Composition of Hardened Paste of Soda Residue-Slag-Cement Binding Material System. *Front. Mater.* **2019**, *6*, 211. [CrossRef]
13. Zhao, X.; Liu, C.; Zuo, L.; Wang, L.; Zhu, Q.; Liu, Y.; Zhou, B. Synthesis and characterization of fly ash geopolymer paste for goaf backfill: Reuse of soda residue. *J. Clean. Prod.* **2020**, *260*, 121045. [CrossRef]
14. Wang, Q.; Li, J.; Yao, G.; Zhu, X.; Arulrajah, A.; Hu, S.; Qiu, J.; Chen, P.; Lyu, X. Characterization of the mechanical properties and microcosmic mechanism of Portland cement prepared with soda residue. *Constr. Build. Mater.* **2020**, *241*, 117994. [CrossRef]
15. Bai, X.; Ma, J.; Liu, J.; Zhang, M.; Yan, N.; Wang, Y. Field experimental investigation on filling the soda residue soil with liquid soda residue and liquid fly ash. *Int. J. Damage Mech.* **2020**, *30*, 502–517. [CrossRef]
16. Ma, J.; Yan, N.; Zhang, M.; Liu, J.; Bai, X.; Wang, Y. Mechanical Characteristics of Soda Residue Soil Incorporating Different Admixture: Reuse of Soda Residue. *Sustainability* **2020**, *12*, 5852. [CrossRef]
17. Yan, Y.; Sun, X.; Ma, F.; Li, J.; Shen, J.; Han, W.; Liu, X.; Wang, L. Removal of phosphate from wastewater using alkaline residue. *J. Environ. Sci.* **2014**, *26*, 970–980. [CrossRef]
18. Ucal, G.O.; Mahyar, M.; Tokyay, M. Hydration of alinite cement produced from soda waste sludge. *Constr. Build. Mater.* **2018**, *164*, 178–184. [CrossRef]
19. Liu, J.Z.; Zhao, Q.X.; Zhang, J.R.; An, S. Microstructure and composition of hardened paste of soda residue-slag complex binding materials. *J. Build. Mater.* **2019**, *22*, 872–877.
20. Lin, Y.H.; Xu, D.Q.; Zhao, X.H. Experimental research on mechanical property and microstructure of blast furnace slag cementitious materials activated by soda residue. *Bull. Chin. Ceram. Soc.* **2019**, *38*, 2876–2881.
21. Guo, W.; Zhang, Z.; Bai, Y.; Zhao, G.; Sang, Z.; Zhao, Q. Development and characterization of a new multi-strength level binder system using soda residue-carbide slag as composite activator. *Constr. Build. Mater.* **2021**, *291*, 123367. [CrossRef]
22. Guo, W.; Wang, S.; Xu, Z.; Zhang, Z.; Zhang, C.; Bai, Y.; Zhao, Q. Mechanical performance and microstructure improvement of soda residue-carbide slag-ground granulated blast furnace slag binder by optimizing its preparation process and curing method. *Constr. Build. Mater.* **2021**, *302*, 124403. [CrossRef]
23. Li, J.; Zhang, S.; Wang, Q.; Ni, W.; Li, Z. Feasibility of using fly ash–slag-based binder for mine backfilling and its associated leaching risks. *J. Hazard. Mater.* **2020**, *400*, 123191. [CrossRef]
24. Guo, W.; Zhang, Z.; Zhao, Q.; Song, R.; Liu, J. Mechanical properties and microstructure of binding material using slag-fly ash synergistically activated by wet-basis soda residue-carbide slag. *Constr. Build. Mater.* **2021**, *269*, 121301. [CrossRef]
25. GB/T 17671; Method of Testing Cements -Determination of Strength (ISO method). Standards Press of China: Beijing, China, 1999.
26. GB/T 2419; Test Method for Fluidity of Cement Mortar. Standards Press of China: Beijing, China, 2005.
27. Zhang, J.; Scherer, G.W. Comparison of methods for arresting hydration of cement. *Cem. Concr. Res.* **2011**, *41*, 1024–1036. [CrossRef]
28. Jiao, Z.Z.; Wang, Y.; Zheng, W.Z.; Huang, W.X. Effect of dosage of sodium carbonate on the strength and drying shrinkage of sodium hydroxide based alkali-activated slag paste. *Constr. Build. Mater.* **2018**, *179*, 11–24. [CrossRef]
29. Li, W.; Yi, Y. Use of carbide slag from acetylene industry for activation of ground granulated blast-furnace slag. *Constr. Build. Mater.* **2020**, *238*, 117713. [CrossRef]
30. Zheng, W.; Zou, M.; Wang, Y. Literature review of alkali-activated cementitious materials. *J. Build. Struct.* **2019**, *40*, 28–39.
31. Salih, M.A.; Farzadnia, N.; Ali, A.A.; Demirboga, R. Effect of different curing temperatures on alkali activated palm oil fuel ash paste. *Constr. Build. Mater.* **2015**, *94*, 116–125. [CrossRef]
32. Lin, Y.; Xu, D.; Zhao, X. Properties and hydration mechanism of soda residue-activated ground granulated blast furnace slag cementitious materials. *Materials* **2021**, *14*, 2883. [CrossRef]
33. Xu, D.; Ni, W.; Wang, Q.; Xu, C.; Jiang, Y. Preparation of clinker-free concrete by using soda residue composite cementitious material. *J. Harbin Inst. Technol.* **2020**, *52*, 151–160.
34. Jiao, Z.Z.; Wang, Y.; Zheng, W.Z.; Huang, W.X. Effect of the activator on the performance of alkali-activated slag mortars with pottery sand as fine aggregate. *Constr. Build. Mater.* **2019**, *197*, 83–90. [CrossRef]
35. Wang, W.C.; Wang, H.Y.; Lo, M.H. The fresh and engineering properties of alkali activated slag as a function of fly ash replacement and alkali concentration. *Constr. Build. Mater.* **2015**, *84*, 224–229. [CrossRef]
36. Taylor, H.; Famy, C.; Scrivener, K. Delayed ettringite formation. *Cem. Concr. Res.* **2001**, *31*, 683–693. [CrossRef]

37. Tian, X.; Xu, W.Y.; Song, S.X.; Rao, F.; Xia, L. Effects of curing temperature on the compressive strength and microstructure of copper tailing-based geopolymers. *Chemosphere* **2020**, *253*, 126754. [CrossRef]
38. Phoo-Ngernkham, T.; Phiangphimai, C.; Intarabut, D.; Hanjitsuwan, S.; Damrongwiriyanupap, N.; Li, L.-Y.; Chindaprasirt, P. Low cost and sustainable repair material made from alkali-activated high-calcium fly ash with calcium carbide residue. *Constr. Build. Mater.* **2020**, *247*, 118543. [CrossRef]
39. Han, F.; Liu, R.; Yan, P. Influence of slag on microstructure of complex binder pastes. *J. Chin. Electron Microsc. Soc.* **2014**, *33*, 40–45.
40. Zhang, J.; Tan, H.; He, X.; Yang, W.; Deng, X. Utilization of carbide slag-granulated blast furnace slag system by wet grinding as low carbon cementitious materials. *Constr. Build. Mater.* **2020**, *249*, 118763. [CrossRef]
41. Park, H.; Jeong, Y.; Jun, Y.; Jeong, K.-H.; Oh, J.E. Strength enhancement and poresize refifinement in clinker-free CaO-activated GGBFS systems through substitution with gypsum. *Cem. Concr. Compos.* **2016**, *68*, 57–65. [CrossRef]
42. Ping, Y.; Kirkpatrick, R.J.; Poe, B.; Mcmillan, P.F.; Cong, X. Structure of Calcium Silicate Hydrate (C-S-H): Near-, Mid-, and Far-Infrared Spectroscopy. *J. Am. Ceram. Soc.* **2010**, *82*, 742–748.
43. Mollah, M.Y.A.; Yu, W.; Schennach, R.; Cocke, D.L. A Fourier transform infrared spectroscopic investigation of the early hydration of Portland cement and the influence of sodium lignosulfonate. *Cem. Concr. Res.* **2000**, *30*, 267–273. [CrossRef]
44. Ghosh, S.N.; Handoo, S.K. Infrared and Raman spectral studies in cement and concrete (review). *Cem. Concr. Res.* **1980**, *10*, 771–782. [CrossRef]
45. Lodeiro, I.G.; Macphee, D.E.; Palomo, A.; Fernández-Jiménez, A. Effect of alkalis on fresh C–S–H gels. FTIR analysis. *Cem. Concr. Res.* **2009**, *39*, 147–153. [CrossRef]
46. Fernandez, L.; Alonso, C.; Hidalgo, A.; Andrade, C. The role of magnesium during the hydration of C3S and CSH formation. Scanning electron microscopy and mid-infrared studies. *Adv. Cem. Res.* **2005**, *17*, 9–21. [CrossRef]
47. Zhang, Y.; Sun, W.; Jia, Y.; Jin, Z. Composition and structure of hardened geopolymer products using infrared ray analysis methods. *J. Wuhan Univ. Technol.* **2005**, *27*, 31–34.
48. Puertas, F.; Palacios, M.; Manzano, H.; Dolado, J.; Rico, A.; Rodríguez, J. A model for the CASH gel formed in alkali-activated slag cements. *J. Eur. Ceram. Soc.* **2011**, *31*, 2043–2056. [CrossRef]
49. Zhang, Y.; Liu, X.; Xu, Y.; Tang, B.; Wang, Y.; Mukiza, E. Synergic effects of electrolytic manganese residue-red mud-carbide slag on the road base strength and durability properties. *Constr. Build. Mater.* **2019**, *220*, 364–374.
50. Liu, C.B.; Ji, H.G.; Liu, J.H.; He, W.; Gao, C. Experimental study on slag composite cementitious material for solidifying coastal saline soil. *J. Build. Mater.* **2015**, *18*, 82–87.

Article

Optimizing the Mechanical Performance and Microstructure of Alkali-Activated Soda Residue-Slag Composite Cementing Materials by Various Curing Methods

Zhaoyun Zhang [1,2,*], Chuang Xie [1,*], Zhaohu Sang [2] and Dejun Li [2]

1 School of Chemical Engineering and Technology, Tianjin University, Tianjin 300072, China
2 Tangshan Sanyou Alkali Chloride Co., Ltd., Tangshan 063000, China
* Correspondence: zhang924001@163.com (Z.Z.); acxie@tju.edu.cn (C.X.)

Abstract: Aiming to promote further the application of alkali-activated soda residue-ground granulated blast furnace slag (SR-GGBS) cementing materials, this study explored the optimal curing method for enhancing mechanical performance. The optimal curing method was determined based on the development of compressive strengths at different curing periods and microstructural examination by XRD, FTIR, SEM, and TG-DTG. The results show that the strength of cementing materials after room-temperature (RT) dry curing was the poorest, with the slow development of mechanical performance. The 7d and 28d compressive strengths were only 14.62 and 20.99 MPa, respectively. Compared with the values after RT dry curing, the samples' 7d and 28d compressive strengths after RT water curing, standard curing, and RT sealed curing were enhanced by 16.35%/24.06%, 30.98%/23.77%, and 38.24%/37.97%, respectively. High-temperature (HT) curing can significantly improve the early strength of the prepared cementing materials. Curing at 60 °C for 12 h was the optimal HT curing method. Curing at 60 °C for 12 h enhanced the 3d strength by 100.84% compared with standard curing. This is because HT curing promoted the decomposition and aggregation of GGBS, and more C-A-S-H gel and crystal hydration products, including ettringite and calcium chloroaluminate hydrate, were produced and filled the inner pores, thereby enhancing both the overall compactness and mechanical performance. However, curing at too high temperatures for too long can reduce the material's overall mechanical performance. After excess HT curing, many shrinkage cracks were produced in the sample. Different thermal expansion coefficients of different materials led to a decline in strength. The present study can provide a theoretical foundation for extensive engineering applications of alkali-activated SR-GGBS composite cementing materials.

Keywords: composite cementing materials; soda residue; ground granulated blast furnace slag; mechanical performance; curing method

1. Introduction

Alkali-activated cementing materials are hydraulic cementing materials prepared under the action of alkali activators based on catalysis principles [1]. Over the past two decades, alkali-activated cementing materials attracted extensive interest worldwide. Compared with Portland cement, alkali-activated cementing materials possess several advantages, including low energy consumption, high intensity, and good durability in production and performance [2,3]. The main hydration products of Portland cement are hydrated calcium silicate and calcium hydroxide. In contrast, alkali-containing aluminosilicate gels with poor crystallinity act as the main products of alkali-activated cementing materials, which can also account for their excellent durability [4]. In addition to industrial byproducts, alkali-activated materials can also be used for waste disposal and environmental protection [5].

Many cementing materials, including ground granulated blast furnace slag (GGBS), steel slag [6], and fly ash [7], can be used for preparing alkali-activated materials. Alkali-activated GGBS cementing materials have received the most extensive research. Nedunuri

Citation: Zhang, Z.; Xie, C.; Sang, Z.; Li, D. Optimizing the Mechanical Performance and Microstructure of Alkali-Activated Soda Residue-Slag Composite Cementing Materials by Various Curing Methods. *Sustainability* 2022, 14, 13661. https://doi.org/10.3390/su142013661

Academic Editor: Ning Yuan

Received: 28 September 2022
Accepted: 20 October 2022
Published: 21 October 2022

Publisher's Note: MDPI stays neutral with regard to jurisdictional claims in published maps and institutional affiliations.

Copyright: © 2022 by the authors. Licensee MDPI, Basel, Switzerland. This article is an open access article distributed under the terms and conditions of the Creative Commons Attribution (CC BY) license (https://creativecommons.org/licenses/by/4.0/).

et al., investigated the mechanical and chemical structural changes of alkali-activated GGBS cementing materials in the early stage [8]. It was found that the setting time of the cementing material drops with the increasing addition of GGBS, accompanied by an increase in hydration products and hydration heat. As the Moore modulus and the mixing amount of activator increase, the setting time can be shortened. The rapid setting of the alkali-activated GGBS mixture refers to the rapid setting of the gel network accompanied by the formation of hydration products. Lv et al., investigated the applicability of seawater to the preparation of GGBS-fly ash alkali-activated materials. They found that adding seawater can accelerate the alkali activation process, shorten the setting time and enhance both early and long-term strength [9]. According to their results, the 3d and 1-year compressive strengths exceeded 25 and 73 MPa, respectively. Moreover, adding seawater negatively affected rebars and expanded the application range of alkali-activated materials. Adesanya et al. used desulfurization dust (DeS-dust) to replace sodium hydroxide as an alkali activator to activate GGBS cementing materials [10]. Using DeS-dust as an alkali activator, the 28d compressive strength can reach up to 33 MPa; in contrast, the 28d compressive strength of the materials after the addition of sodium hydroxide was only 25 MPa. Based on the micromeasurement results, the microstructures after the addition of the two activators are comparable, and the cementing materials with DeS-dust exceeded those with sodium hydroxide in performance.

At normal temperature, because of the high activity of GGBS, alkali-activated cementing materials undergo hydration under an alkali activator with a specific strength [11,12]. However, to save cost, many solid wastes, such as calcium carbide slag [13] and gypsum [14], have been added to alkali-activated GGBS cementing materials in recent years. Considering the low activity of wastes at normal temperatures, the curing of wastes should be optimized to achieve better performance. Guo et al. [15] used calcium carbide slag slurry and soda residue (SR) slurry to activate GGBS and prepare a new kind of cementing material; according to their experimental results, the 3d and 28d compressive strengths of the prepared samples can be enhanced by 50.0 and 34.7%, respectively, compared with the samples produced with the addition of powders. It was concluded that curing at 60 °C for 12 h proved to be optimal. Compared with the samples after water curing, the 3d compressive strength of the sample after curing at 60 °C for 12 h can be enhanced by 66.7%. This is because high temperature (HT) promotes the decomposition and aggregation of GGBS, thereby accelerating the generation of C-S-H gel and calcium chloroaluminate hydrate crystals. However, after curing at too high a temperature for a long time, the long-term strength of the prepared sample drops, which can be attributed to the formation of drying shrinkage cracks in the sample. By adding NaOH, Na_2SiO_3, and silicon manganese dust (SMF) to slag (BFS), Nasir et al. [16] prepared a kind of alkali-activated cementing material and determined the optimal curing conditions in the oven (at 60 °C for 6 h); after curing at 60 °C for 6 h, the 3d, 7d, and 28d compressive strengths were 38, 41, and 45.2 MPa, respectively. As the crystallinity degree of the C-(A)-S-H and quartz phases increased, stratlingite and gehlenite were formed to fill the internal pores and form special phases, such as C-Mn-S-H and K-A-S-H.

SR is a kind of alkali waste residue discharged during the production of sodium carbonate with the ammonia-soda process. SR is characterized by strong alkalinity, with a pH value of approximately 11–12, and mainly consists of $CaCO_3$, $Ca(OH)_2$, $CaCl_2$, and $CaSO_4$ [17–19]. In particular, China has a vast production scale of sodium carbonate, with an annual SR discharge of over 1000 t. Approximately 0.3 t SR should be discharged for preparing 1 t sodium carbonate [20]. SR accumulation can bring about high processing costs for enterprises and occupy a large land area. The harmful ingredients are easily permeated into the soil, which can cause land salinization and lower soil quality. Moreover, SR powder particles are quite fine and can be easily inhaled by people, thereby endangering health [21]. In recent years, some scholars have conducted an increasing number of studies on preparing a novel kind of alkali-activated cementing material based on the characteristics of SR, which can solve the secondary pollution problems caused by SR. An et al. used gypsum powder-SR-GGBS to prepare a novel cementing material, in which the mixing ratio of gypsum powder was 0, 5, and 10%, the mixing ratios of SR were 0, 70, and 80%, and the rest was GGBS [22]. The results show that the strength dropped with the increasing addition of SR. The optimal mixing ratios of gypsum powder and SR were determined to be 5% and 60%, respectively. Under the optimal conditions, the 28d unconfined compressive strength was 9310 kPa. Because of the activation of SR, C-S-H gel was produced, and AFt acted as the main crystal product and provided the primary source of strength. Lin et al. employed strong alkaline SR to activate GGBS and measured the prepared mortar samples' fluidity and compressive strength. According to their results, under the condition (with a water-to-binder ratio of 0.5, SR and GGBS mixing ratios of 16–24% and 76–84%, respectively. The 28d compressive strength was 32.3–35.4 MPa, and the fluidity ranged from 181 to 195 mm. The main hydration products were ettringite, Friedel's salt, and C-S-H gels.

Finally, based on previous research results [3] and a literature review, this study aimed to optimize the curing method of alkali-activated SR-GGBS cementing materials. The most appropriate curing methods can be obtained by observing the strength development and microformation process of the different samples. The present research results can contribute to promoting and popularizing the SR-GGBS system.

2. Experimental Section

2.1. Materials

The raw materials used in this study mainly include SR, GGBS, NaOH, standard sand, and water. The chemical components, micromorphology, and mineral composition of SR and GGBS were measured via XRF, SEM, and XRD, and the results are illustrated in Figures 1 and 2, as well as tabulated in Table 1.

(a) (b)

Figure 1. SEM images of (a) SR and (b) GGBS.

Table 1. Chemical composition of GGBS and SR.

Material	Chemical Composition (wt/%)/%											
	CaO	SiO$_2$	Al$_2$O$_3$	Fe$_2$O$_3$	MgO	TiO$_2$	K$_2$O	SO$_3$	MnO	Na$_2$O	Cl$^-$	LOI *
GGBS	33.70	32.60	17.10	1.18	7.96	2.54	0.57	3.21	0.31	0.56	-	2.14
SR	43.20	9.87	3.25	0.91	9.77	0.12	0.29	5.57	-	3.93	23.00	2.86

* LOI = loss on ignition.

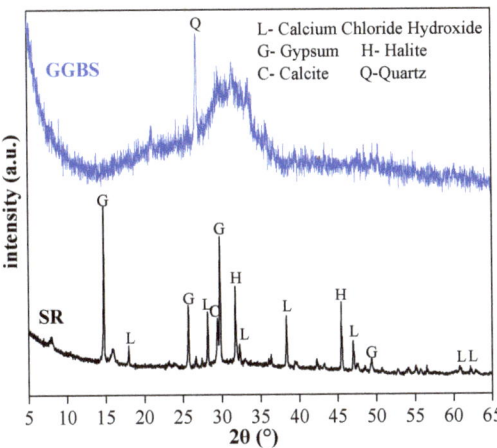

Figure 2. XRD spectra of SR and GGBS.

The particle sizes of the two raw materials were analyzed with a BT-9300H Baxter laser particle sizer, and the results are shown in Figure 3. The specific surface area was measured via the gas absorption BET method, and the density was measured using Lee's bottle method, as shown in Table 2.

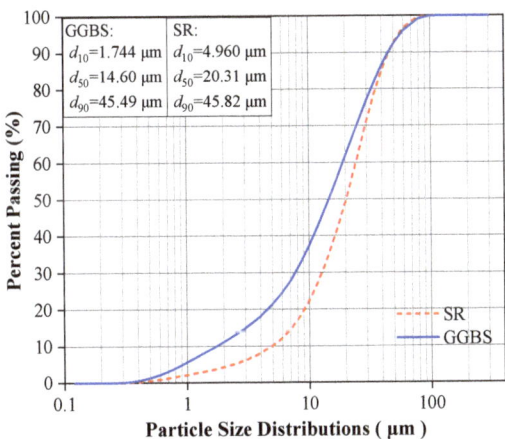

Figure 3. Particle size distributions of SR and GGBS.

Table 2. Physical properties of GGBS and SR.

Material	Physical Properties	
	Specific Gravity (g/cm^2)	Specific Surface Area (m^2/kg)
GGBS	2.742	419.5
SR	2.351	261.2

This study used SR manufactured by Tangshan Sanyou Chlor-Alkali Co., Ltd., Hebei, China. The initial moisture content of the SR mortar exceeded 90%. Before use, the slurry was allowed to stand still for 2–8 days, and the upper clear liquid was removed until the slurry was separated. The lower slurry with a moisture content of 50–60% was dried for

24 h and ground to SR powder with a particle size below 0.16 mm. As shown in Figure 1a, SR particles could be easily aggregated with many pores. The SR mineral phases were mainly calcium chloride hydroxide, gypsum, halite, and calcite. According to the XRF measurement results, SR is primarily composed of CaO (43.2%) and Cl^- (23.0%), with small fractions of SiO_2 (9.87%) and Al_2O_3. The density, specific surface area, and mean particle size of SR were 2742 kg/m^3, 419.5 m^2/kg, and 20.31 μm, respectively. GGBS, from Hebei, China, was ground with S95 GGBS. As shown in Figure 1b, GGBS was composed of irregular blocks in terms of micromorphology. The main mineral phase of GGBS was quartz. Based on XRF measurement results, GGBS under study was composed of CaO, SiO_2, and Al_2O_3 in terms of chemical components, with proportions of 33.7, 32.6, and 17.1%, respectively. The density, specific surface area, and mean particle size of SR were 2742 kg/m^3, 419.5 m^2/kg, and 14.6 μm, respectively. The sand used in this study satisfied the GB/T17671 standard, with a particle size range of 0.08–2.0 mm and a sediment content below 0.2%. Tap water in the laboratory was also used in this study.

2.2. Mixing Ratios

Based on previous research results, the SR-to-GGBS ratio was fixed at 4:6. This study focused on the effect of different curing methods (including room temperature (RT) water curing, RT dry curing, standard curing, RT sealed curing, and HT curing on the strength and microstructure of prepared novel gel samples. Table 3 lists the detailed mixing ratios and curing methods.

Table 3. Samples, mixing ratios, and curing methods used in this study.

Sample ID	Dosage of Dry Basis/g		Sand/g	Water-to-Binder Ratio	Curing Method
	SR	GGBS			
F1					RT *, Water curing
F2					RT *, Dry curing
F3					Standard curing
F4					RT *, Sealed curing
F5					60 °C @ 6 h
F6	180	270	1350	0.5	60 °C @ 12 h
F7					60 °C @ 24 h
F8					40 °C @ 12 h
F9					50 °C @ 12 h
F10					70 °C @ 12 h
F11					80 °C @12 h

* RT = room temperature, 20 ± 2 °C.

2.3. Sample Preparation

Before the test, water usage was first calculated based on the preset water-to-binder ratio. NaOH particles were first dissolved in water and stood still for 24 h. The detailed preparation procedures are described below. First, NaOH solution was poured into the agitating pan, and the mixture of SR powder and GGBS was added. The pan was fixed on the frame and lifted to the operating position. Second, the machine started for 1 min of stirring at low speed. The neat paste remained on the vanes, and the pan wall was spaded into the pan for 2 min of high-speed stirring to prepare neat paste samples. Third, the samples were stirred at low speed for 30 s, and standard sand was added. Then, the samples were stirred at high speed for 1.5 min to prepare the mortar samples. Finally, after adequate stirring, the samples were hierarchically put into test molds with exact sizes of 40 × 40 × 160 mm and then vibrated for 120 s. After the test mold was taken down, the sample was leveled and numbered.

2.4. Curing Procedures

The samples were demolded 24 h after pouring and then cured by methods listed in Table 3. The F1 sample was placed into the curing tank at RT (20 ± 2 °C) and cured to the preset period. The F2 sample was cured at RT to the preset period. The F3 sample was put into the standard curing box at a temperature of (20 ± 2 °C) and humidity of over 90% and cured for the specified period. The F4 sample was wrapped with plastic film and cured at RT to the preset period. Samples F5–F11 were wrapped into plastic film, placed into an oven for HT curing, then transferred to RT for further sealed curing until the specified period.

2.5. Testing Procedure

The compressive strengths were measured according to the GB/T17671 standard. The test periods were set as 3, 7, and 28 days. At each age, the samples were measured six times for averaging. For each sample, the mean and standard deviation were used for characterization. The load was uniformly applied to the sample during the test at a loading rate of 2400 N/s ± 200 N/s until failure.

Paste samples with a cubic size of 50 mm × 50 mm × 50 mm were used in this study. After curing to the preset period according to the methods described in Table 3, the sample was loaded to failure. The central region without carbonization was taken and soaked in isopropanol for 3–7 days to avoid further hydration. Finally, the small blocks were removed and placed into a vacuum-drying oven to remove the residual isopropanol. Small samples with dry and smooth surfaces and maximum side lengths below 10 mm were selected for the SEM test. Before XRD, FTIR, and TD-DTG tests, small dry samples were removed and ground into a powder with a particle size below 0.075 mm by an agate mortar. The powder sample was then placed into a vacuum bag for standby application. During the present XRD test, a Cu target was used; the tube voltage and current were set as 40 kV and 40 mA, respectively; the scanning angle ranged from 5 to 65°, with the scanning rate preset at 2 °/min. The wavenumber in the present FTIR test varied within a range of 4000–400 cm^{-1} at a resolution of 2 cm^{-1}. During the TG-DTG test, the temperature varied from RT to 1000 °C at a heating rate of 10 °C/min, and Ar acted as the protective gas. XRD, FTIR, SEM, and TG-DTG were tested using a D/MAX-2500/PC X-ray diffractometer, BRUKER TENSOR II infrared spectrometer, TESCAN-VEGA3, and STA449C/6/G integrated thermal analyzer, respectively.

3. Results and Discussion

3.1. Compressive Strength

Figure 4 displays the measured results of the samples' 3d, 7d, and 28d compressive strengths after different curing methods, curing temperatures, and curing times.

3.1.1. Effect of the Curing Method on the Compressive Strength

Figure 4a depicts the effect of different curing methods on compressive strength. The compressive strength increased with the prolonging of the curing period. The 3d compressive strengths of the samples after water curing, RT dry curing, standard curing, and RT sealed curing were 10.62, 11.03, 9.55, and 9.83 MPa, respectively, with slight differences. After drying at 60 °C for 12 h (first curing at 60 °C for 12 h and then under standard curing conditions to the specified period), the 3d compressive strength could reach up to 19.18 MPa. This might be because appropriate HT curing accelerated the hydration process, generating more hydration products, and enhancing structural compactness, thereby improving the sample's early strength. As the curing period increased to 7 and 28 days, the sample after RT dry curing showed the poorest performance and slow mechanical performance development. For the sample after RT dry curing, the 7d and 28d compressive strengths were 14.62 and 20.09 MPa, respectively. The reason was that water could be rapidly evaporated in a dry environment, leading to insufficient water consumption for hydration and incomplete hydration. Therefore, the sample has low strength [15]. In contrast to samples after RT dry curing, the samples' 7d and 28d compressive strengths after water curing, standard curing, and RT sealed curing increased significantly, amounting to 17.01/26.04 MPa, 19.15/25.98 MPa, and 20.21/28.96 MPa, respectively. It can be seen that the strength of RT sealed curing in the 3 and 7 days is slightly different from the standard curing, and it is only slightly higher than the standard curing in the 28 days, which indicates that the water supply of this new type of binder under sealed curing can meet the long-term hydration demand, and does not need high humidity, which may also be an advantage of this binder. The 7d and 28d compressive strengths of the mortar sample after HT curing were the highest, reaching 25.37 and 30.11 MPa, respectively.

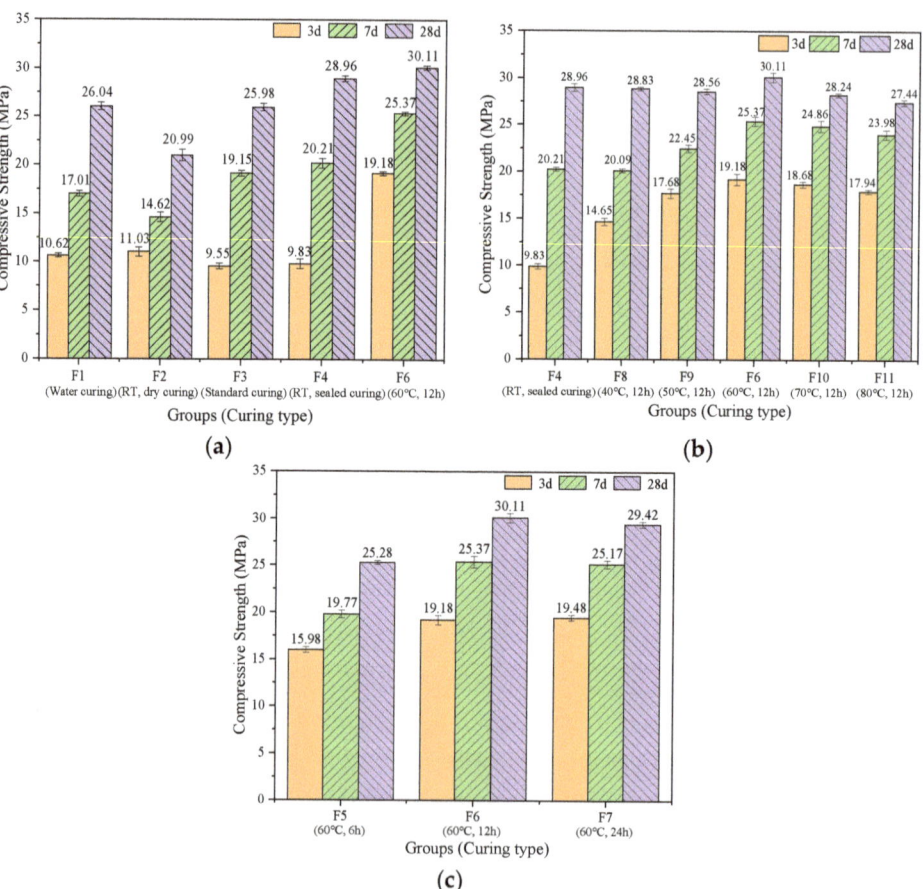

Figure 4. Test results on the compressive strength: (**a**) effect of curing methods; (**b**) effect of HT curing temperature; (**c**) effect of HT curing time.

3.1.2. Effects of High-Temperature Curing Temperature on Compressive Strength

Figure 4b shows the effect of the curing temperature on the compressive strength. After curing at high temperatures (40, 50, 60, 70, and 80 °C) for 12 h, the samples were cured to the specified periods under standard curing conditions. It can be found that different curing temperatures impose different effects on the development of compressive strength. For sealed curing from RT to 60 °C, the compressive strength was positively correlated with the curing temperature. With increasing temperature, the 3d, 7d, and 28d compressive strengths increased from 9.83, 20.21, and 28.96 MPa to 17.44, 25.37, and 30.11 MPa, respectively. As the curing temperature increased, the compressive strength was negatively correlated with the curing temperature, accompanied by decreases in the 3d, 7d, and 28d compressive strengths from 18.68, 24.86, and 28.24 MPa to 17.94, 23.98, and 27.44 MPa. Accordingly, the optimal curing temperature was determined to be 60 °C. Previous experimental studies have reported similar findings [15,16]. In particular, Guo et al. revealed that curing at 60 °C enhanced the early strength of alkali-activated GGBS [15]. The 3d compressive strength after curing at 60 °C was 19.81 MPa. The 3d compressive strengths of the samples after curing at RT, 40, 50, 70, and 80°C were 51.25, 76.38, 97.39, and 93.53% of the values after curing at 60 °C. The 7d and 28d compressive strengths after curing at 60 °C were 25.37 and 30.11 MPa, respectively. Similarly, compared with the values after curing at 60 °C, the samples' 7d and 28d compressive strengths after curing at room temperature, 40, 50, 70, and 80 °C were 79.66%/96.18%, 79.19%/95.75%, 88.49%/94.85%, 97.98%/93.79%, and 94.52%/91.13%, respectively. This suggests that curing at an appropriate high temperature can enhance the early strength and impose no adverse effect on the long-term strength.

3.1.3. Effect of High-Temperature Curing Time on Compressive Strength

Figure 4c illustrates the effect of different HT curing times on compressive strength. The samples were first cured at 60 °C for 6, 12, and 24 h, then placed at RT and cured for the specified periods. As the curing time increased from 6 to 12 and 24 h, the 3d compressive strength increased steadily from 15.98 and 19.18 MPa to 19.48 MPa. The 7d and 28d compressive strengths increased and then dropped, as shown in Figure 4. By considering the strength and energy consumption cost, curing at 60 °C for 12 h was the optimal curing time. As the curing time increased to 24 h, the strength dropped, confirming the adverse effect of long-term HT curing on the strength. This is consistent with previous research results. After heating at a high temperature for a long time, cracks were quickly produced. This can be attributed to different thermal expansion coefficients among different raw materials. Long-term HT curing can lead to the formation of cracks, thereby deteriorating the strength.

3.2. X-ray Diffraction (XRD) Analysis

The produced hydration products were analyzed via XRD based on macromechanical test results to investigate further the difference in mechanical performance under different curing conditions. The 3d and 28d XRD patterns under different curing conditions (RT sealed curing, curing at 60 °C for 12 h, curing at 60 °C for 24 h, and curing at 80 °C for 12 h) are displayed in Figure 5 for comparison. As expected, the sample after HT curing showed a noticeable change.

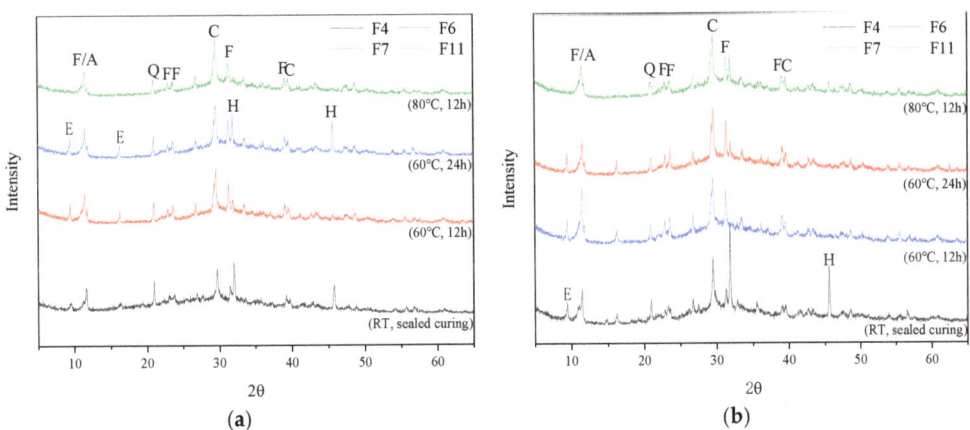

Figure 5. XRD patterns of samples: (**a**) 3 days; (**b**) 28 days. Hereinafter, the following designations are used: A = hydrotalcite, F = calcium chloroaluminate hydrate, C = calcite, H = halite, Q = quartz, E = ettringite.

According to Figure 5a, the 3d crystal hydration products after the RT sealed curing were ettringite (E) and calcium chloroaluminate hydrate (F). Specifically, quartz (Q) was sourced from GGBS, and calcite (C) was sourced from SR. The F-phase and E-phase were enhanced by comparing the patterns of the F4 and F6 samples, while the quartz peak dropped. This suggests that curing at an appropriate temperature can promote the hydration of GGBS and the production of C-A-S-H gel, as well as E and F crystals, thereby enhancing the early strength.

By comparing the patterns of the F6 and F7 samples, the peaks of the E and F crystals showed no changes with prolonged HT curing time. The characteristic peak of quartz dropped, suggesting constant hydration of GGBS. By comparing the XRD patterns of the F6 and F11 samples, the peak of E disappeared with increased curing temperature. This is because E is greatly affected by temperature. At too high a temperature, E may be decomposed into AFm. Accordingly, no diffraction peak of E was observed at 80 °C [23,24]. The decrease in the F and A peaks suggests that too high a temperature was unfavorable for forming F and A.

As shown in Figure 5b, compared with the 3d XRD patterns of the samples, F and E of the F4, F6, and F7 samples were enhanced, while the peak of quartz dropped. This suggests that hydration of GGBS occurred steadily. F, A, and E were gradually produced during the hydration process. No E peaks appeared in the 28d XRD pattern of the F11 sample, indicating the thorough decomposition of E. In addition, humps can be observed in the XRD patterns of all samples at 25–35°, suggesting the production of amorphous C-A-S-H gel in the hydration products [25]. For the samples after curing at 60 °C for 12 and 24 h, both the height and width of the hump were maximal, suggesting that an appropriate curing temperature can promote the generation of gel and the development of strength.

Overall, curing at an appropriate high temperature can promote the dissolution of GGBS in an alkaline environment and generate more hydration products for gap filling. Accordingly, the sample's mechanical performance can be enhanced. This finding is consistent with the measured compressive strength results.

3.3. FTIR Analysis

Figure 6 displays the FTIR spectra of the F4, F6, F7, and F11 samples after curing periods of 3 and 28 days.

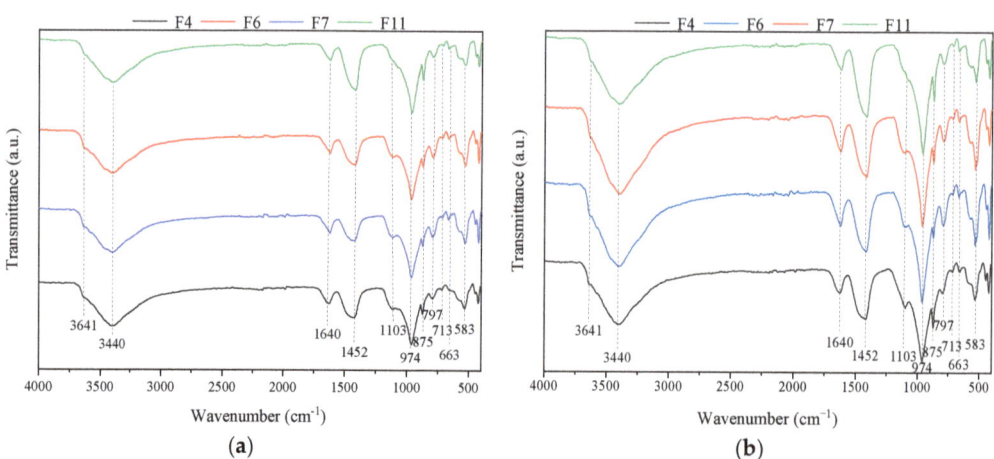

Figure 6. FTIR curves of samples after curing periods of 3d (**a**) and 28d (**b**).

The peak at 583 cm^{-1} can be attributed to the asymmetric stretching vibration of Si-O-Al, and the peaks at 797 and 974 cm^{-1} to the stretching vibration of Si-O. All these absorption peaks are induced by forming a C-A-S-H gel [26,27]. The peak at 663 cm^{-1} can be attributed to the vibration of Mg-O and Al-O [25]. The absorption peak at 1103 cm^{-1} can be attributed to the stretching vibration of SO_4^{2-} [28]. The peaks at 713, 875, and 1452 cm^{-1} correspond to the asymmetric structures of CO_3^{2-} [29,30]. The peaks at 1640 and 3641 cm^{-1} can be attributed to the bending vibration and stretching vibration of the H-O-H bond in crystal water. The peak at 3440 cm^{-1} corresponds to the stretching vibration of Al-OH stretching vibration in the octahedral structure of $[Al(OH)_6]^{3-}$, implying the formation of F [31,32].

By comparing the FTIR patterns of the F4 and F6 samples, the peaks at 583, 797, and 974 cm^{-1} increased after curing at 60 °C for 12 h, suggesting that appropriate curing can promote the formation of C-A-S-H. The peaks at 663, 1103, and 3440 cm^{-1} were enhanced, indicating that HT curing could accelerate the formation of E, F, and A. This is consistent with the present XRD measured results. As the curing time increased from 12 to 24 h, the peaks corresponding to C-A-S-H increased, but the peaks corresponding to A, F, and E varied slightly. Accordingly, it can be concluded that curing time can promote gel formation but imposes a slight effect on the production of crystal hydration products. However, based on the research results by Guo et al., a curing time that is too long may lead to the formation of fractures [15]. As the curing temperature increased, the peak at 1103 cm^{-1} disappeared, indicating the decomposition of F at high temperatures. This fits well with the XRD measurement results that the peaks of the C-A-S-H gel, F, and A crystals varied slightly.

As the curing period increased to 28 days, the characteristic peaks at 583, 663, 797, 974, 1103, and 3440 cm^{-1} depicted in Figure 6b were enhanced compared with the peaks in the FTIR curves of 3d samples in Figure 6a. Accordingly, increasing amounts of A, F, and E were generated, suggesting that hydration was a sustainable development process. The observations from FTIR curves fit well with the XRD results.

3.4. SEM Analysis

To characterize the effects of different curing methods, temperatures, and times on the sample's microstructure, SEM images of the F4, F6, F7, and F11 samples at a curing period of 3 days are displayed in Figure 7. Under RT sealed curing (as shown in Figure 7a), unreacted GGBS, SR, and a few C-A-S-H gels can be observed. The produced gel wrapped unreacted GGBS. The needle-shaped E was interwoven into a network. Overall, the structure was relatively loose with low compactness. Many pores can be observed, which can be attributed to the slow reaction of alkali-activated GGBS. Figure 7b depicts the results of the samples after curing at 60 °C for 12 h. The microstructure can be improved to a certain degree, and only a few unreacted particles were observed, while an increasing amount of E was produced. A large amount of C-A-S-H gel was produced and filled the pores, thereby improving structural compactness. This suggests that appropriate HT curing can accelerate the activity of hydroxide ions in alkali solution, accelerating the solution rate of aluminum silicate precursors [33].

This can also account for the favorable mechanical properties of alkali-activated GGBS after curing at 60 °C. After curing at 60 °C from 12 to 24 h, many fractures can be observed on the surface despite the generation of many gels. This is consistent with the research results by Nasir [16]. As the curing temperature increases to 80 °C, almost no E can be observed, suggesting that E has almost been decomposed. This is consistent with the above-described XRD results. At excessively high temperatures, the surface of the raw materials was wrapped by the produced hydration products, which can inhibit the reaction with the alkali solution, thereby reducing the long-term strength. According to the present experimental results, 60 °C was determined as the optimal curing temperature.

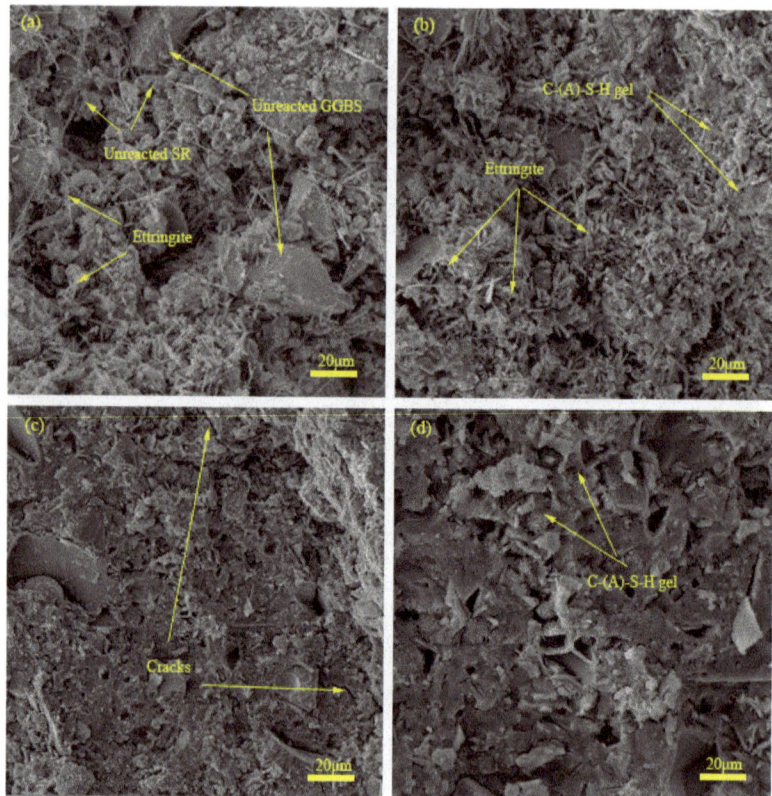

Figure 7. SEM images of the samples after a 3d curing period: (**a**) F4, (**b**) F6, (**c**) F7, and (**d**) F11.

3.5. Thermogravimetry/Derivative Thermogravimetry Analysis (TG-DTG)

Thermogravimetry/derivative thermogravimetry (TG-DTG) curves of the F4, F6, F7, and F11 samples after a curing period of 3 days are plotted in Figure 8, while Figure 9 presents the statistics on weight losses of four samples in a temperature range from 50 to 400 °C. In Figure 8, three weight loss peaks can be observed, which are located at 50–200 °C, 250–400 °C, and 400–600 °C.

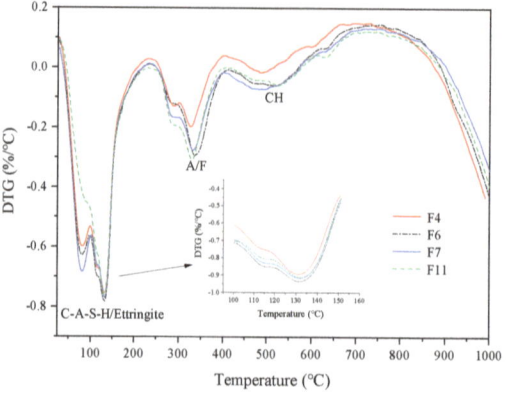

Figure 8. DTG curves of 3d−cured specimens.

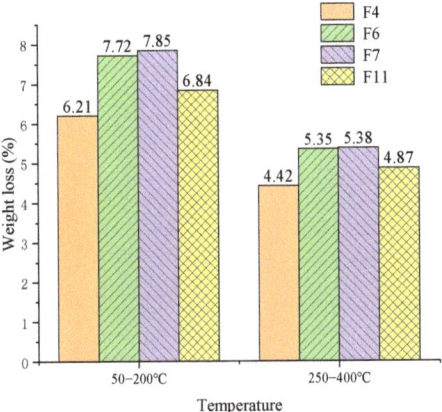

Figure 9. Weight loss of 3d−cured specimens in different temperature ranges.

The peak at 50–200 °C can be attributed to the water loss-induced decomposition of the C-A-S-H gel and E. Specifically, the peak at 85–110 °C corresponds to E [34]. Compared with the condition under RT curing, the water loss peak was enhanced after heating, accompanied by an increase in the weight loss ratio from 6.21 to 7.72%. Accordingly, it can be concluded that curing at an appropriate temperature can promote the formation of E and C-A-S-H gels. The generated hydration products filled the pores. Therefore, the sample can obtain higher mechanical performance, which is consistent with the measured results of compressive strength. With prolonged curing time, the water loss ratio increased steadily from 7.22 to 7.85%, suggesting the constant formation of hydration products. As the curing temperature rose to 80 °C, the characteristic peak can be attributed to the formation of the C-A-S-H gel because of the disappearance of E. Meanwhile, the water loss ratio was reduced to 6.84%. This suggests that curing at too high temperatures can reduce the production of hydration products. On the one hand, high temperature can promote the decomposition of E, reducing its further formation. On the other hand, the surface of raw materials was wrapped by the hydration products rapidly generated at high temperatures, hindering further reaction and thereby reducing the strength [35].

The peak at 250–400 °C was induced by A and F [36,37]. For the four samples, the weight loss ratios were 4.42, 5.35, 5.38, and 4.87%, respectively. As the curing temperature increased from RT to 60 °C, the weight loss ratio increased, suggesting that curing at an appropriate temperature can promote the formation of crystal hydration products such as F and A. As the curing time increased, the weight loss ratio varied slightly, suggesting that curing at too a high temperature imposed almost no effect on the production of crystal hydration products such as A and F. As the curing temperature further increased, the weight loss ratio dropped. Thus, an excessively high curing temperature has an adverse effect on the production of crystal hydration products, including A and F.

The peak at 400–600 °C seems to be mainly related to water loss of $Ca(OH)_2$ [16]. However, the XRD and FTIR analyses revealed no apparent changes in $Ca(OH)_2$ under different curing conditions, which may be due to the less amount of $Ca(OH)_2$ generated.

4. Conclusions

This study analyzed the effects of different curing methods on the mechanical performance and microstructure of alkali-activated SR-GGBS cementing materials. The experimental results obtained made it possible to draw the following conclusions:

(1) For alkali-activated SR-GGBS cementing samples under different curing conditions, the sample after RT sealed curing showed optimal mechanical performance. The compressive strength increased from 9.83 MPa at 3d to 28.96 MPa at 28d. However,

(2) In contrast to the above four curing methods, HT curing at an appropriate temperature can remarkably enhance the early strength of the prepared alkali-activated SR-GGBS cementing materials without harming their long-term strength. This is because the hydration of GGBS is promoted at HT, generating more C-A-S-H gel and crystal hydration products, which fill the internal pores and improve the material integrity.

(3) HT curing at 60 °C for 12 h proved to be the optimal method among the investigated ones. Noteworthy is that excessive HT curing deteriorates long-term strength, producing shrinkage cracks and hindering further strength development.

(4) HT curing at 60 °C for 12 h is the most appropriate for preparing samples with high early strength in rapid construction. For the samples with a low requirement on early strength, RT sealed curing is suitable for reduced energy consumption and convenient operation advantages.

Author Contributions: Conceptualization, Z.Z.; Formal analysis, C.X. and D.L.; Funding acquisition, C.X.; Investigation, Z.S.; Methodology, Z.Z.; Project administration, Z.S.; Writing–original draft, Z.Z. and Z.S.; Writing–review & editing, C.X. and D.L. All authors have read and agreed to the published version of the manuscript.

Funding: This research was funded by the National Natural Science Foundation of China under the grant of 21776204.

Acknowledgments: Appreciated the Tangshan Sanyou Alkali Chloride Co., Ltd., for technical consulting, and Tianjin University for academic assist.

Conflicts of Interest: The authors declare no conflict of interest.

References

1. Guo, W.; Bai, Y.; Xu, Z.; Zhang, J.; Zhao, Q.; Wang, D. Stress-strain behavior of low-carbon concrete activated by soda residue-calcium carbide slag under uniaxial and triaxial compression. *J. Build. Eng.* **2022**, *55*, 104678. [CrossRef]
2. Guo, W.; Zhang, Z.; Xu, Z.; Zhang, J.; Bai, Y.; Zhao, Q.; Qiu, Y. Mechanical properties and compressive constitutive relation of solid waste-based concrete activated by soda residue-carbide slag. *Constr. Build. Mater.* **2022**, *333*, 127352. [CrossRef]
3. Zhang, Z.; Xie, C.; Sang, Z.; Li, D. Mechanical Properties and Microstructure of Alkali-Activated Soda Residue-Blast Furnace Slag Composite Binder. *Sustainability* **2022**, *14*, 11751. [CrossRef]
4. Athira, V.; Bahurudeen, A.; Saljas, M.; Jayachandran, K. Influence of different curing methods on mechanical and durability properties of alkali activated binders. *Constr. Build. Mater.* **2021**, *299*, 123963. [CrossRef]
5. Celikten, S.; Saridemir, M.; Deneme, I.O. Mechanical and microstructural properties of alkali-activated slag and slag plus fly ash mortars exposed to high temperature. *Constr. Build. Mater.* **2019**, *217*, 50–61. [CrossRef]
6. Liu, Z.; Zhang, D.-W.; Li, L.; Wang, J.-X.; Shao, N.-N.; Wang, D.-M. Microstructure and phase evolution of alkali-activated steel slag during early age. *Constr. Build. Mater.* **2019**, *204*, 158–165. [CrossRef]
7. Siddique, S.; Jang, J.G. Acid and sulfate resistance of seawater based alkali activated fly ash: A sustainable and durable approach. *Constr. Build. Mater.* **2021**, *281*, 122601. [CrossRef]
8. Nedunuri, A.S.S.S.; Muhammad, S. Fundamental understanding of the setting behaviour of the alkali activated binders based on ground granulated blast furnace slag and fly ash. *Constr. Build. Mater.* **2021**, *291*, 123243. [CrossRef]
9. Lv, W.; Sun, Z.; Su, Z. Study of seawater mixed one-part alkali activated GGBFS-fly ash. *Cem. Concr. Compos.* **2020**, *106*, 103484. [CrossRef]
10. Adesanya, E.; Ohenoja, K.; Di Maria, A.; Kinnunen, P.; Illikainen, M. Alternative alkali-activator from steel-making waste for one-part alkali-activated slag. *J. Clean. Prod.* **2020**, *274*, 123020. [CrossRef]
11. Jiao, Z.; Wang, Y.; Zheng, W.; Huang, W. Effect of the activator on the performance of alkali-activated slag mortars with pottery sand as fine aggregate. *Constr. Build. Mater.* **2019**, *197*, 83–90. [CrossRef]
12. Bernal, S.A.; DE Gutierrez, R.M.; Provis, J. Engineering and durability properties of concretes based on alkali-activated granulated blast furnace slag/metakaolin blends. *Constr. Build. Mater.* **2012**, *33*, 99–108. [CrossRef]
13. Kou, R.; Guo, M.-Z.; Han, L.; Li, J.-S.; Li, B.; Chu, H.; Jiang, L.; Wang, L.; Jin, W.; Poon, C.S. Recycling sediment, calcium carbide slag and ground granulated blast-furnace slag into novel and sustainable cementitious binder for production of eco-friendly mortar. *Constr. Build. Mater.* **2021**, *305*, 124772. [CrossRef]
14. Gao, D.; Zhang, Z.; Meng, Y.; Tang, J.; Yang, L. Effect of Flue Gas Desulfurization Gypsum on the Properties of Calcium Sulfoaluminate Cement Blended with Ground Granulated Blast Furnace Slag. *Materials* **2021**, *14*, 382. [CrossRef]

15. Guo, W.; Wang, S.; Xu, Z.; Zhang, Z.; Zhang, C.; Bai, Y.; Zhao, Q. Mechanical performance and microstructure improvement of soda residue–carbide slag–ground granulated blast furnace slag binder by optimizing its preparation process and curing method. *Constr. Build. Mater.* **2021**, *302*, 124403. [CrossRef]
16. Nasir, M.; Johari, M.A.M.; Maslehuddin, M.; Yusuf, M.O.; Al-Harthi, M.A. Influence of heat curing period and temperature on the strength of silico-manganese fume-blast furnace slag-based alkali-activated mortar. *Constr. Build. Mater.* **2020**, *251*, 118961. [CrossRef]
17. Zhao, X.; Liu, C.; Wang, L.; Zuo, L.; Zhu, Q.; Ma, W. Physical and mechanical properties and micro characteristics of fly ash-based geopolymers incorporating soda residue. *Cem. Concr. Compos.* **2019**, *98*, 125–136. [CrossRef]
18. Song, R.; Zhao, Q.; Zhang, J.; Liu, J. Microstructure and Composition of Hardened Paste of Soda Residue-Slag-Cement Binding Material System. *Front. Mater.* **2019**, *6*, 211. [CrossRef]
19. Guo, W.; Zhang, Z.; Bai, Y.; Zhao, G.; Sang, Z.; Zhao, Q. Development and characterization of a new multi-strength level binder system using soda residue-carbide slag as composite activator. *Constr. Build. Mater.* **2021**, *291*, 123367. [CrossRef]
20. Zhao, X.; Liu, C.; Zuo, L.; Wang, L.; Zhu, Q.; Liu, Y.; Zhou, B. Synthesis and characterization of fly ash geopolymer paste for goaf backfill: Reuse of soda residue. *J. Clean. Prod.* **2020**, *260*, 121045. [CrossRef]
21. Wang, Q.; Li, J.; Yao, G.; Zhu, X.; Hu, S.; Qiu, J.; Chen, P.; Lyu, X. Characterization of the mechanical properties and microcosmic mechanism of Portland cement prepared with soda residue. *Constr. Build. Mater.* **2020**, *241*, 117994. [CrossRef]
22. An, Q.; Pan, H.; Zhao, Q.; Du, S.; Wang, D. Strength development and microstructure of recycled gypsum-soda residue-GGBS based geopolymer. *Constr. Build. Mater.* **2022**, *331*, 127312. [CrossRef]
23. Lin, Y.; Xu, D.; Ji, W.; Zhao, X. Experiment on the Properties of Soda Residue-Activated Ground Granulated Blast Furnace Slag Mortars with Different Activators. *Materials* **2022**, *15*, 3578. [CrossRef]
24. Christensen, A.N.; Jensen, T.R.; Hanson, J.C. Formation of ettringite, $Ca_6Al_2(SO_4)_3(OH)_{12}$ center dot $26H_2O$, AFt, and monosulfate, $Ca_4Al_2O_6(SO_4)$ center dot $14H_2O$, AFm-14, in hydrothermal hydration of Portland cement and of calcium aluminum oxide-calcium sulfate dihydrate mixtures studied by in situ synchrotron X-ray powder diffraction. *J. Solid State Chem.* **2004**, *177*, 1944–1951.
25. Li, W.; Yi, Y. Use of carbide slag from acetylene industry for activation of ground granulated blast-furnace slag. *Constr. Build. Mater.* **2020**, *238*, 117713. [CrossRef]
26. Yu, P.; Kirkpatrick, R.J.; Poe, B.; McMillan, P.F.; Cong, X. Structure of Calcium Silicate Hydrate (C-S-H): Near-, Mid-, and Far-Infrared Spectroscopy. *J. Am. Ceram. Soc.* **1999**, *82*, 742–748. [CrossRef]
27. Mollah, M.Y.A.; Yu, W.; Schennach, R.; Cocke, D.L. A Fourier transform infrared spectroscopic investigation of the early hydration of Portland cement and the influence of sodium lignosulfonate. *Cem. Concr. Res.* **2000**, *30*, 267–273. [CrossRef]
28. Ghosh, S.N.; Handoo, S.K. Infrared and Raman spectral studies in cement and concrete (review). *Cem. Concr. Res.* **1980**, *10*, 771–782. [CrossRef]
29. Lodeiro, I.G.; Macphee, D.E.; Palomo, A.; Fernandez-Jimenez, A. Effect of alkalis on fresh C-S-H gels. FTIR Anal. *Cem. Concr. Res.* **2009**, *39*, 147–153. [CrossRef]
30. Fernández, L.; Alonso, C.; Hidalgo, A.; Andrade, C. The role of magnesium during the hydration of C3S and C-S-H formation. Scanning electron microscopy and mid-infrared studies. *Adv. Cem. Res.* **2005**, *17*, 9–21. [CrossRef]
31. Zhang, Y.; Sun, W.; Jia, Y.; Jin, Z. Composition and structure of hardened geopolymer products using infrared ray analysis methods. *J. Wuhan Univ. Technol.* **2005**, *27*, 31–34.
32. Liu, X.; Zhao, X.; Yin, H.; Chen, J.; Zhang, N. Intermediate-calcium based cementitious materials prepared by MSWI fly ash and other solid wastes: Hydration characteristics and heavy metals solidification behavior. *J. Hazard. Mater.* **2018**, *349*, 262–271. [CrossRef]
33. Rovnaník, P. Effect of curing temperature on the development of hard structure of metakaolin-based geopolymer. *Constr. Build. Mater.* **2010**, *24*, 1176–1183. [CrossRef]
34. Abdalqader, A.F.; Jin, F.; Al-Tabbaa, A. Development of greener alkali-activated cement: Utilisation of sodium carbonate for activating slag and fly ash mixtures. *J. Clean. Prod.* **2016**, *113*, 66–75. [CrossRef]
35. Schöler, A.; Lothenbach, B.; Winnefeld, F.; Zajac, M. Hydration of quaternary Portland cement blends containing blast-furnace slag, siliceous fly ash and limestone powder. *Cem. Concr. Compos.* **2015**, *55*, 374–382. [CrossRef]
36. Bai, Y.; Guo, W.; Wang, J.; Xu, Z.; Wang, S.; Zhao, Q.; Zhou, J. Geopolymer bricks prepared by MSWI fly ash and other solid wastes: Moulding pressure and curing method optimisation. *Chemosphere* **2022**, *307*, 135987. [CrossRef]
37. Zhang, J.; Tan, H.; He, X.; Yang, W.; Deng, X. Utilization of carbide slag-granulated blast furnace slag system by wet grinding as low carbon cementitious materials. *Constr. Build. Mater.* **2020**, *249*, 118763. [CrossRef]

Article

Significant Fragmentation of Disposable Surgical Masks—Enormous Source for Problematic Micro/Nanoplastics Pollution in the Environment

Alen Erjavec *, Olivija Plohl, Lidija Fras Zemljič and Julija Volmajer Valh

Institute of Engineering Materials and Design, Faculty of Mechanical Engineering, University of Maribor, Smetanova 17, SI-2000 Maribor, Slovenia
* Correspondence: alen.erjavec@um.si

Abstract: The pandemic of COVID-19 disease has brought many challenges in the field of personal protective equipment. The amount of disposable surgical masks (DSMs) consumed increased dramatically, and much of it was improperly disposed of, i.e., it entered the environment. For this reason, it is crucial to accurately analyze the waste and identify all the hazards it poses. Therefore, in the present work, a DSM was disassembled, and gravimetric analysis of representative DSM waste was performed, along with detailed infrared spectroscopy of the individual parts and in-depth analysis of the waste. Due to the potential water contamination by micro/nanoplastics and also by other harmful components of DSMs generated during the leaching and photodegradation process, the xenon test and toxicity characteristic leaching procedure were used to analyze and evaluate the leaching of micro/nanoplastics. Micro/nanoplastic particles were leached from all five components of the mask in an aqueous medium. Exposed to natural conditions, a DSM loses up to 30% of its mass in just 1 month, while micro/nanoplastic particles are formed by the process of photodegradation. Improperly treated DSMs pose a potential hazardous risk to the environment due to the release of micro/nanoparticles and chloride ion content.

Keywords: DSM; micro/nanoparticles; leaching; artificial weathering; environmental pollution

Citation: Erjavec, A.; Plohl, O.; Zemljič Fras, L.; Valh Volmajer, J Significant Fragmentation of Disposable Surgical Masks—Enormous Source for Problematic Micro/Nanoplastics Pollution in the Environment. *Sustainability* 2022, *14*, 12625. https://doi.org/10.3390/su141912625

Academic Editor: Ning Yuan

Received: 12 July 2022
Accepted: 30 September 2022
Published: 4 October 2022

Publisher's Note: MDPI stays neutral with regard to jurisdictional claims in published maps and institutional affiliations.

Copyright: © 2022 by the authors. Licensee MDPI, Basel, Switzerland. This article is an open access article distributed under the terms and conditions of the Creative Commons Attribution (CC BY) license (https://creativecommons.org/licenses/by/4.0/).

1. Introduction

Since the discovery of Bakelite in 1907 until now, the plastic industry has undergone exceptional development and has had a great impact on our daily life [1]. According to Plastics Europe, global plastic production is increasing enormously year by year. For example, 367 million tonnes of plastic were produced in 2020. Plastics Europe estimates that at least half of all plastics produced have a short lifespan [2].

With the outbreak of COVID-19 and the announcement of a pandemic on 11 March 2020 by the World Health Organization (WHO), the global demand for medical personal protective equipment (PPE) has increased [3]. As Park and colleagues note in their analysis, global production of PPE, a short lifespan product, would need to increase by 40% during the pandemic to meet crisis demand [4]. Prevention of human-to-human transmission of the virus SARS-CoV-2 has led to the worldwide consumption of disposable surgical masks (DSMs). The WHO estimates that 89 million medical masks are needed each month to deal with COVID-19 [3]. As Czigany and Ronkay note in their article, surgical masks provide protection and an effective way to keep the virus from circulating because they reduce the number of droplets that an infected person spreads in their environment. Therefore and consequently, protective masks were one of the most sought-after products last year [5].

With the increase in production and consumption of surgical masks in the world, new challenges are being posed to the environment due to the world's largest issue now—microplastics and, even worse, nanoplastics. The declaration of an epidemic due to COVID-19 has triggered a different kind of emergency: single-use plastic is on the rise again worldwide, and much of the non-recyclable PPE is consequently disposed of in the environment causing potential chemical contamination [6] and potential ecotoxicological consequences [7,8].

Generally, DSMs are classified as "FFP1" masks according to the EU standard EN149 and are also known by the code N95, as they can achieve 95% filtration of particles with a diameter of 0.3 µm. They are composed of five parts [9] that include ear loops, nose wire, and three layers of microfibers or nanofibers, which are hydrophobic, skin-friendly, and non-allergenic (Figure 1). The filter layers are usually produced using melt-blown electrospinning technology. All three layers can be made from a variety of synthetic polymeric materials such as polypropylene, polyurethane, polyacrylonitrile, polystyrene, polycarbonate, polyethylene, or polyester, depending on the customer's preference [5,7]. However, as it is a plastic product, it is expected that after usage, in the form of waste, it will have further negative impact on the already polluted environment [5,9].

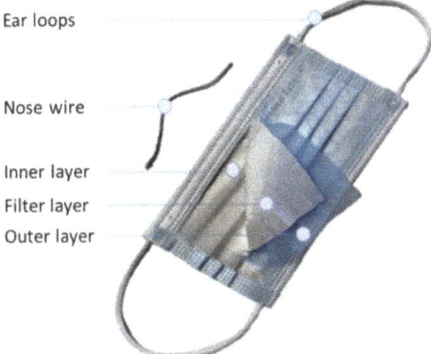

Figure 1. Composition of typical DSM with marked pieces.

Nowadays, DSMs are becoming a new and enormous source of plastic and micro/nanoplastics in the environment and causing severe pollution. For these reasons, it is of paramount importance to study the release of hazardous chemicals and fragmentation products from DSMs. A number of studies have already been published on the above topic, either containing only theoretical investigations on the negative effects of DSMs on the environment or focusing only on new DSMs from one or a few manufacturers. The studies mainly investigated the leaching of DSMs during the washing process or in artificial seawater [5–8,10–18] In the context of aging experiments, Francesco Saliu and others [9] performed an initial and preliminary evaluation by subjecting commercially available surgical masks to artificial aging experiments, including UV irradiation and mechanical stress in artificial seawater. The results showed that a single surgical mask irradiated with UV light for 180 h and vigorously agitated in artificial seawater can release up to 173,000 fibers/day. Silvia Morgana et al. also state that a single mask in water can release thousands of microplastic fibers and up to 10^8 submicrometric particles, most of which are nanoscale [19].

The advantage of our study over other published studies is that we used a real sample—a mixture of used DSMs of different colors and from different manufacturers—to obtain a real representative sample. Our waste samples contained more than 30 different DSMs that are used daily in Slovenia. This is a much more heterogeneous sample than the ones used in similar studies where more homogeneous samples were used (usually less than 10 [8,16] or even only one type of DSM [6,17]). A representative sample consisting of DSMs was collected and a comprehensive study of the waste DSMs using a series of chemical, thermochemical, and physical analyses was conducted to characterize the DSMs as waste. Gravimetric analysis of larger samples (100 masks) and a complete waste characterization has been carried out in accordance with EU legislation.

In our study, special attention was paid to the potential water pollution from micro/nanoplastics generated by the degradation of these masks and also from other harmful components of DSMs generated during the leaching and photodegradation process, highlighting their adverse effects if not disposed of properly. For this purpose, two methods, a xenon arc fading lamp test in accordance with standard ISO 105-B04:1994 and toxicity leaching procedure (TCLP), were used, to analyze the production and characteristics of micro/nanoplastics [8,12,13,16].

In the xenon test chamber, the weathering resistance of the DSM was evaluated using different light spectra, temperatures, and humidity. In this way, the study of weathering and accelerated aging tests is made possible by closely mimicking actual environmental conditions.

2. Materials and Methods

2.1. Sampling and Gravimetric Analysis

The waste DSM sample was collected in a collection campaign in the faculty of the Mechanical Engineering University of Maribor. In the campaign were participating students, professors, and staff employed in the faculty who regularly delivered spent DSMs. During the campaign, over 10.000 spent DSMs were collected. A sampling of the whole collected sample was carried out according to a standard [20] to collect a representative sample for further analysis. Particular caution was given when collecting and sampling, due to the potential infectiveness of the material.

The surgical masks were divided into 5 basic components—nose wire, ear loops, outer non-woven layer, melt-blown filter layer, and inner non-woven layer—and was dried to constant mass. On a sample of 100 disassembled masks, we performed a gravimetric analysis of the proportion of each component in our sample. The masks were disassembled manually. Each component was weighed on a Kern ALT 220-4NM balance with an accuracy of ± 0.001 g.

2.2. Characterization of Physicochemical Properties of Surgical Masks

The DSMs were analyzed with the spectrometer ATR FTIR Perkin Elmer Spectrum GX (Perkin Elmer FTIR, Omega, Ljubljana, Slovenia). The ATR accessory (supplied by Specac Ltd., Orpington, Kent, UK) contained a diamond crystal. A total of 16 scans were taken of each sample with a resolution of 4 cm^{-1}. All spectra were recorded at ambient temperature over a wavenumber interval between 4000 and 650 cm^{-1}. Four measurements were performed for all 5 parts and on both sides of each layer of DSMs. Pristine DSMs were also characterized in terms of thermogravimetric properties and DSC as explained below.

2.3. Waste Characterization

In accordance with the Slovenian regulation [21] that is modeled after Council Directive 1999/31/EC, chemical thermogravimetrical and additional physicochemical analyses were performed to carry out the characterization of waste. First, a sample was prepared in accordance with the standard [22]. The metal wires were removed from the sample and the remainder was ground into fine dust using an A 10 IKA Werke mill with a water-cooling system. Fifty grams of milled sample was produced which is roughly equal to the mass of 100 masks. So, it contained all 3 layers of the mask, ear loops, type 3 nose wire, and type 1 and 2 plastic coating of the nose wire (in accordance with Figure 2). With the obtained sample, the determination of the content of metal elements was carried out using inductively coupled plasma with optical emission spectrometry (ICP-OES) in line with the standard [23]. As in the standard [24], the determination of total organic carbon (TOC) in waste was performed. The nitrogen content was determined according to the standard [25] and hydrogen using the Dumas method [26]. Similar to the standard [27], the determination of loss on ignition was carried out and the dry matter content according to the standard [28] method A. Determination of gross calorific value and calculation of net calorific value were also carried out as in the standard [29]. Other methods that have been used were the determination of polychlorinated biphenyls (PCBs) [30] and the determination of chlorine (Cl), fluorine (F), and sulfur (S) [31].

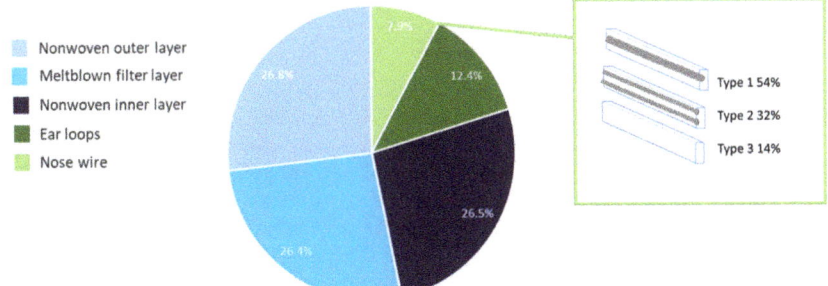

Figure 2. Gravimetric analysis of various DSM compositions shown as a percentage with included types of nose wire and their proportion in our sample.

2.4. Toxicity Characteristic Leaching Procedure (TCLP) and Analysis of TCLP Products

The toxicity characteristic leaching procedure (TCLP)—SW-846 Test Method 1311 [32]— was performed to determine the mobility of both organic and inorganic analytes present in samples. A representative sample of DSMs was prepared in the following way: all 5 parts (nose wire, ear loops and outer, filter, and inner layer) were cut into pieces smaller than 1 cm^2. Then, the dry matter of the sample was determined according to the standard [28]. The TCLP was performed for the whole mask and all 5 parts of the DSM separately. Six different samples were analyzed, and each was prepared in two parallels—outer layer, filter layer, inner layer, ear loops, nose wire separately and all parts together in mass proportions the same as for mask composition. This value was used in Equation (1) to determine the amount of distilled water that is needed to perform a TCLP of 50 g of our sample. Fifty grams of the sample was put in a glass jar with the correct amount of distilled water. Then, the jars were put on an HS-501 shaker made by IKA Werke that rotated at 30 ± 2 rpm for 18 ± 2 h.

$$m_{fluid} = \frac{20 \cdot w_{dry\ matter} \cdot m_{sample}}{100} \quad (1)$$

Products were first filtered through a colander with a mesh having a diameter of 5 mm, to remove larger parts. Then, obtained filtrate was filtered again through polyether sulfone (PESU) filters with an effective pore size of 0.2 μm and diameter of 50 mm, from Sartorius Stedim Biotech GmbH. Microplastics retained on the filter were examined under a Zeiss Axio Vision optical microscope with which different parts were observed and measured. Dried microplastics on the filter were also weighed and analyzed using an ATR FTIR Perkin Elmer Spectrum GX spectrometer as is described in Section 2.2. Filters used were dried to constant mass at 100 °C in a VS50-SC dryer produced by Kambič d.o.o. and weighed before the filtration. After filtration, they were dried again and weighed to determine the mass of microplastics on the filter. Filters were weighed on a Kern ALT 220-4NM balance with an accuracy of ±0.001 g. Obtained microplastic particles were also analyzed and measured under the optical microscope ZEISS Axiotech 25HD (+ pole), with an AxioCam MRC (D) high-resolution digital camera, and a Carl Zeiss FE-SEM SUPRA 35 VP scanning electron microscope with a GEMINI field emission module.

Water obtained in the TCLP before and after the filtration process was tested for simple qualitative properties. With a Velp Scientifica TB1 turbidimeter, turbidity of samples was measured and, with a Metler-Toledo SevenCompact S230 conductometer, conductivity of samples was measured. Chemical oxygen demand (COD) was also determined by using tube tests by Macherey-Nagel, Macherey-Nagel Nanocolor Vario 4 heating unit, and Macherey-Nagel Nanocolor UV/Vis spectrophotometer.

Water after filtration was also analyzed for particle size distribution (PSD) using a Zetasizer Nano ZS® (Malvern Instruments, Ltd., UK) equipped with dynamic light scattering (DLS) technology. Triplicate measurements were carried out using a He-Ne laser at a wavelength of 633 nm and scattering angle of 173° at 25 °C for 70 s.

2.5. Photodegradability of Mask

Photodegradation of the mask was performed by artificial weathering using a Xenotest alpha LM high-energy climate chamber (Atlas Material Testing Technology GmbH, 1500 Bishop Ct, Mount Prospect, Illinois, 60056, United States). It was essentially the same as the standard ISO 4892-2 2013, with minor modifications in cycle composition. The DSMs were manually prepared to fit into the Xenotest holders. The samples were then dried at 90 °C for 1 h to maintain a constant mass. The samples were then weighed and placed in the Xenotest. One cycle consisted of a 1 min rain period and a 29 min dry period at a relative humidity of 50% and a constant temperature of 38 °C. Irradiance was set to 60 W/m^2 and was provided with the use of a daylight filter. These laboratory conditions are a rapid simulation of the real natural conditions to which DSMs are exposed in nature, thus mimicking real conditions. A cycle was repeated 25 times, 50 times, 75 times, 100 times, 125 times, 150 times, 175 times, and 200 times. Samples were then dried again at 90 °C for 1 h and weighed. In addition to gravimetric analysis, we also performed FTIR analysis as well as thermogravimetric analysis (TGA) and differential scanning calorimetry (DSC) of the masks after aging. Distilled water was used for the simulated rainwater, which was collected and analyzed after the simulation. A few drops of the simulated rainwater were examined under a ZEISS Axiotech 25HD (+ pole) optical microscope. It was also filtered through a PESU filter with < 0.2 μm mesh size. The filtrate was analyzed by DLS to detect any nanoscale particles released from the samples.

TGA and DSC analyses of raw DSMs were also performed to assess possible changes. For TGA before and after Xenotest, a METTLER TOLEDO TGA 2 STAR System was used under air atmosphere in the temperature range 25–700 °C and with a heating rate of 10 K min^{-1}. For DSC, a METTLER TOLEDO DSC 3 STAR System was used under nitrogen (N$_2$) atmosphere in the temperature range of 25–700 °C and with a heating rate of 10 K min^{-1}.

3. Results

3.1. DSM Characterizations

3.1.1. Gravimetric Analysis

After manually decomposing 100 different DSMs into five components (shown in Figure 1), we used the basic statistical tools in Microsoft Excel to obtain the results shown in Figure 2. The average mass of a single mask was 3.1205 g. The maximum mass fraction of the mask represents a three-layered part, resulting in 79.7%. This percentage is evenly distributed among all three layers and most likely contributes to micro/nanoplastic fragmentation. Ear loops represent 12.4% of the mass of the DSMs, and the lowest percentage of the DSM mass is represented by nose wire, i.e., 7.9%.

In the obtained sample, three different types of nose wire were found. The most common one is nose wire composed of one metal wire coated with plastic (type 1) and is represented in more than half of all masks analyzed, the second one is composed of two metal wires coated with plastic (type 2), and the third one, the least represented type, is without metal wire, just plastic with additives that provide similar properties to metal wire in plastic (type 3). Schematic representations of all three types of nose wire are shown in Figure 2.

3.1.2. ATR-FTIR Characterization of Pristine DSMs

Similar to above, 50 different DSMs were manually decomposed into all five parts as shown in Figure 1 and analyzed by ATR FTIR spectroscopy. The results show the different spectra that were collected from each part. As can be seen in Figure 3a, two different types of spectra of the non-woven outer layer and non-woven inner layer have been found. All collected spectra of the filter layer were the same. The spectra of one type of non-woven outer layer, filter layer, and one type of non-woven inner layer showed the following bands: multiple peaks in the wavenumber range from 3000 to 2800 cm^{-1} and two large peaks in the range from 1456 to 1375 cm^{-1}. The peaks in the range from 3000 to 2800 cm^{-1} were attributable to asymmetric and symmetric stretching vibrations of CH_2 groups, while the peaks at 2950 and 2850 cm^{-1} were due to the asymmetric and symmetric stretching vibrations of CH_3. The peak at 1456 cm^{-1} indicates the asymmetric CH_3 vibrations or CH_2 scissor vibrations, while the peak at 1375 cm^{-1} was the result of the symmetric CH_3 deformation [33]. All mentioned signals represent typical signals for polypropylene materials.

The other two spectra, corresponding to the second type of non-woven outer layer and the second type of non-woven inner layer, presented only two signals in the range from 3000 to 2800 cm^{-1}, namely, at the wavenumber 2914 and 2848 cm^{-1}, signal at 1471 cm^{-1}, and the small peak at 717 cm^{-1}, that are identified as characteristic bands for C-H wagging vibrations (Figure 3a). Those signals are attributed to polyethylene materials [34]. Interestingly, among the 50 DSMs measured, only one was found to have a non-woven outer layer and a non-woven inner layer of polyethylene materials.

As can be seen in Figure 3b, four different spectra of ear loops with different positions of signals were obtained. Comparing the collected spectra with the spectra of synthetically produced untreated polyamide 6 [35], it was revealed that the characteristic peaks overlap. Typical bands that were found were assigned to polyamide 6, also known as nylon 6. However, slight variations between the spectra can be observed due to different manufacturers of ear loops adding different additives to provide desired properties of the product.

In the case of nose wire examination, the metal wires were removed first, and then only the plastic coating was analyzed. As already schematically shown in Figure 2, three different types of nose wires were found according to their structure. As they differ in structure, they also differ in the FTIR spectra obtained. Type 1 was a nose wire with a metal wire in a plastic coating. As can be seen in Figure 3c, the spectra of type 1 have peaks typical for polypropylene, and the position of bands is very similar to the spectra of the coatings of polypropylene in Figure 3a. Type 2 and type 3 nose wires' spectra are very similar, and they are compared with the spectra in Figure 3a, as they are also comparable to the inner and outer layer of DSMs made of polyethylene. Indeed, among all nose wires examined, the material polypropylene predominates.

3.2. DSM Characterization as a Waste

To better understand the potential risks posed by discarded DSMs, the total analysis was performed for DSMs in the form of waste by several standards (for details, refer to Section 2.3) to identify the possible release of chemical compounds when a mask is decomposing in the environment, that according to our knowledge has not yet been carried out, but has already been recommended [18]. In accordance with Slovenian regulation [21], the presence of trace elements was analyzed. The results of the detected items are shown in Table 1. All results are in milligrams from a kilogram of dry matter. Results of other analytical methods that were carried out are shown in Table 2.

The elemental analysis that was carried out showed that 73.99% of a waste DSM without metal wires is composed of total carbon (TC), 16.43% of nitrogen (N), and 6.95% of hydrogen (H). In the same sample, 0.63% chlorine (Cl) was also detected (Table 2) which is of particular concern. It should be noted that all results are calculated on dry matter. One of the possibilities is that chlorine gets into the masks during the sterilization process. As Van Loon and colleagues note, two processes that contain chlorine-containing chemical compounds also appear in the review of possible DSM sterilization processes. These are sterilization with bleach (sodium hypochlorite solution) and chlorine dioxide gas (ClO_2). The latter process is much more commonly used in practice and is very likely the reason for the presence of Cl in waste DSMs.

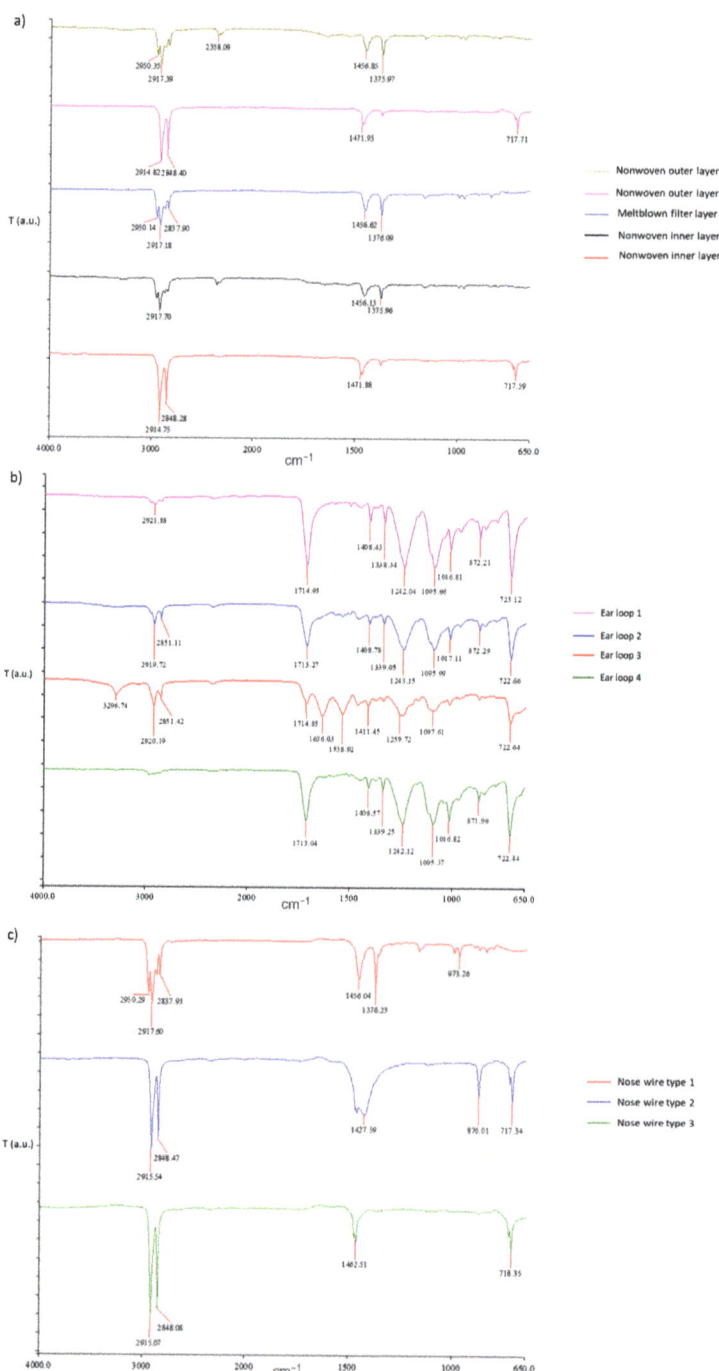

Figure 3. (a) FTIR spectra of 3-layered part of DSM; (b) FTIR spectra of different ear loops; (c) FTIR spectra of nose wires.

Table 1. Analyzed trace elements in waste DSM without metal wire.

Trace Elements	Value (mg/kg d.m.)	Trace Elements	Value (mg/kg d.m.)
Ag	<10	In	<10
Al	39.35	Li	<10
As	<3	Mg	78.41
Ba	<7	Mn	<8
Bi	<10	Na	50.04
Cd	<3	Ni	7.65
Co	<7	Pb	<5
Cr	13.09	Se	<3
Cu	15.42	Sr	<10
Ga	<10	Ti	<10
Hg	<1	Zn	51.97

Table 2. Other analyses to determine waste characterization.

Other Analysis	Value
Dry matter	>99%
Loss on ignition	>99%
Net calorific value	44,264 kJ/kg
Gross calorific value	45,739 kJ/kg
PCB	Not detected
Cl	0.63% d.m.
F	<0.2 mg/kg
S	<0.2 mg/kg

Based on the analysis of the metal content (Table 1), we can conclude that there is no detectable increased presence of metals in our sample. We should keep in mind that we removed the metal wires from the nose wire and analyzed only the polymer coating in the DSM mix sample. Since we provided analyses in accordance with the assessment of hazardous waste prior to incineration [21], there are no limit values. Yet, DSMs are expected to release elements (Table 1) when discarded that can contribute to the release of potentially hazardous chemicals such as Cr, with possible adverse ecological effects on wildlife, as already pointed out [7,15].

As is seen in the data in Table 2, this is a type of waste that is eminently suitable for energy use, as it has an extremely high dry matter content, while at the same time we can conclude from the loss on ignition that no significant amount of ash (less than 1%) is produced during combustion. The net calorific value is also extremely high, even exceeding the net calorific value of crude oil (42,300 kJ/kg) and being comparable to gasoline (≈44,300 kJ/kg) [36].

3.3. Toxicity Characteristic Leaching Procedure (TCLP) and Analysis of TCLP Products and Microplastic That Is Being Produced

Qualitative properties of the water and mass of isolated microplastics obtained in the TCLP are reported in Table 3. During the leaching process, most of the microplastic is released from the ear loops, followed by the inner layer, the outer layer, the filter layer, and the wire, in that order. This is reflected not only in the mass of microplastics larger than 0.2 µm left on the filter but also in the highest COD, turbidity, and conductivity of the washout fluid. These findings were also confirmed by gravimetric analysis of microplastic parts loaded on filters. The greatest mass of microplastic parts is released out of ear loops (28.9 mg) which is 0.23% of the original mass needed for the TCLP procedure. Table 3 shows that the turbidity of each DSM part decreased after filtration. A similar trend is also observed for COD, being smaller after TCLP filtration in all cases. Similarly, conductivity is lower or does not change after filtration. The release of microplastics on PESU filters increased after the TCLP in all parts of the DSM. It can be concluded that turbidity, COD, and conductivity values decrease due to removal of microplastics with the filtration process and that the presence of microplastics increases the values of these three water quality parameters.

Table 3. Simple qualitative properties of the water obtained in TCLP before and after filtration.

DSM Part	Before Filtration			After Filtration			$m_{microplastics>0.2\ \mu m}$ (mg) Released on Filters
	Turbidity (NTU)	Conductivity (µS/cm)	COD (mg/L O_2)	Turbidity (NTU)	Conductivity (µS/cm)	COD (mg/L O_2)	
Blank	0.3	0.1	5.1	0.1	0.1	2.2	0
Outer layer	19.1	90.8	52.5	1.7	89.2	40.5	9.1
Filter layer	9.5	117.2	41.1	0.3	55.5	27.2	5.5
Inner layer	24.9	123.7	151.2	3.8	124.4	124.5	12.6
Ear loops	44.9	204.6	162.4	3.7	199.4	158.0	28.9
Nose wire	6.6	21.7	13.2	0.2	22.5	10.3	2
Whole mask	15.5	103.4	68.3	1.9	111.2	59.3	7.4

The microplastics that were being produced and were loaded on a PESU filter with an effective pore size of 0.2 µm were examined under the optical microscope and with ATR-FTIR analysis. Results of ATR-FTIR analyses of leached microplastic particles by every part of DSM are shown in Figure 4. Pictures made with an optical microscope are shown in Figure 5.

The FTIR spectra of the microplastic produced during the TCLP differ from the FTIR spectra of the base parts of the surgical masks (refer to Figure 3). The FTIR spectra of microplastic release from the outer, inner, and filter layers (Figure 4a–c) and also from the nose wire (Figure 4e) are not consistent with the spectrum of polypropylene, showing new peaks around 3300 cm^{-1}, 1650 cm^{-1} and peaks in the range of 1030–1060 cm^{-1}. The peaks in the range from 3000 to 2800 cm^{-1}, which are due to asymmetric and symmetric stretching vibrations of the CH_2 and CH_3 groups of polypropylenes, are less pronounced and the peaks at 1456 and 1376 cm^{-1} disappear completely in the spectra of the released microplastic. From this, we can conclude that polypropylene undergoes oxidation/degradation reactions that were confirmed with the appearance of new bands after the TCLP [16,18]. The new peak around 3300 cm^{-1} may correspond to hydroxyl groups, while the peaks in the range of 1030–1060 cm^{-1} may represent signals for C-O stretching vibrations [16]. The FTIR spectrum of microplastic released from ear loops (Figure 4d) differs from the polyamide 6 spectrum in the following two peaks: peak at 3295 and peak at 1635 cm^{-1}. The latter may also be attributed to the oxidation process of polymeric material [16,18].

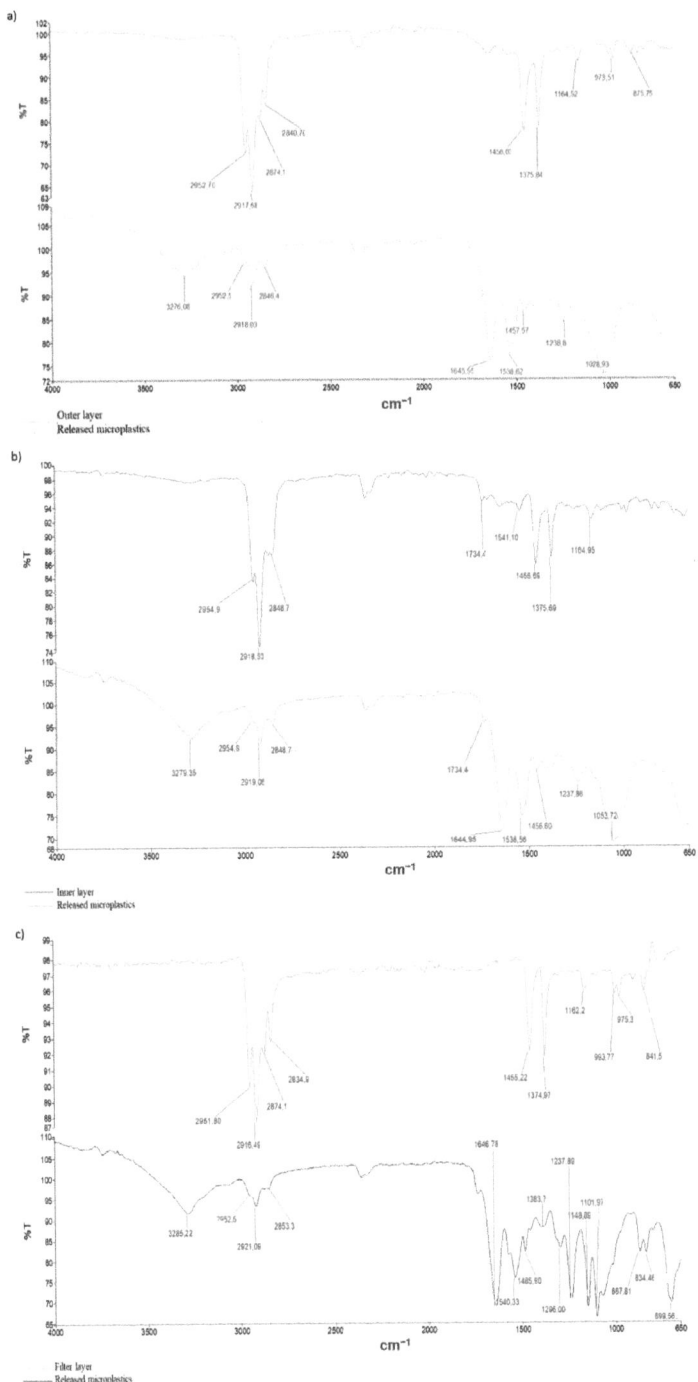

Figure 4. *Cont.*

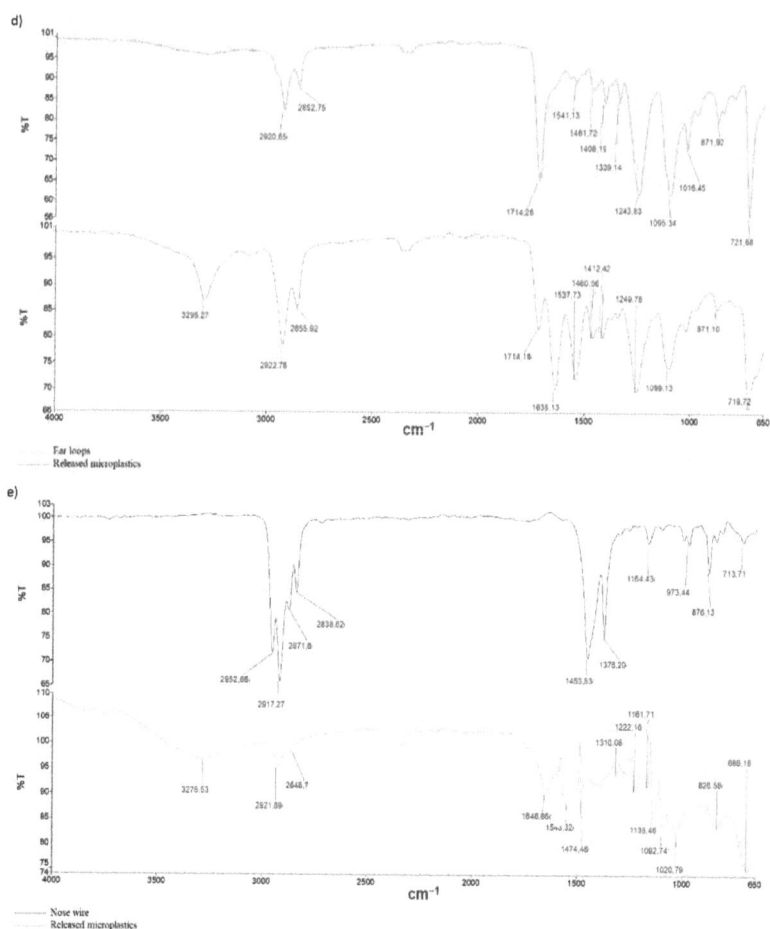

Figure 4. ATR FTIR scans of leached microplastic parts loaded on PESU filter with an effective pore size of 0.2 μm with referential scans of each part before TCLP: (**a**) microplastic parts out of the outer layer, (**b**) microplastic parts out of the inner layer, (**c**) microplastic parts out of filter layer (**d**) microplastic parts out of ear loops, and (**e**) microplastic parts out of nose wire.

The microplastic particles in Figure 5b,d,f,h are quite similar. Most of the particles are fibers that are uniformly distributed and not superimposed. Some fibers are fragmented into particles and broken, which look like irregular spheres and cubes and are no longer fiber-like, as initial layers. Such morphology and shape of fiber-forming particles are expected because of the material from which they are leached (Figure 5a,c,e,g). In Figure 5j, the leached microparticles are different from those in the previous photographs. In this case, the particles are dislodged from nose wires (Figure 5i) and are made of hard polypropylene. The slightly orange color is probably the result of the presence of metal wire. It was shown in [15] that when other particle shapes and morphologies were compared to fiber-shaped particles, it was revealed that the latter in general tends to cause larger ecotoxicological effects.

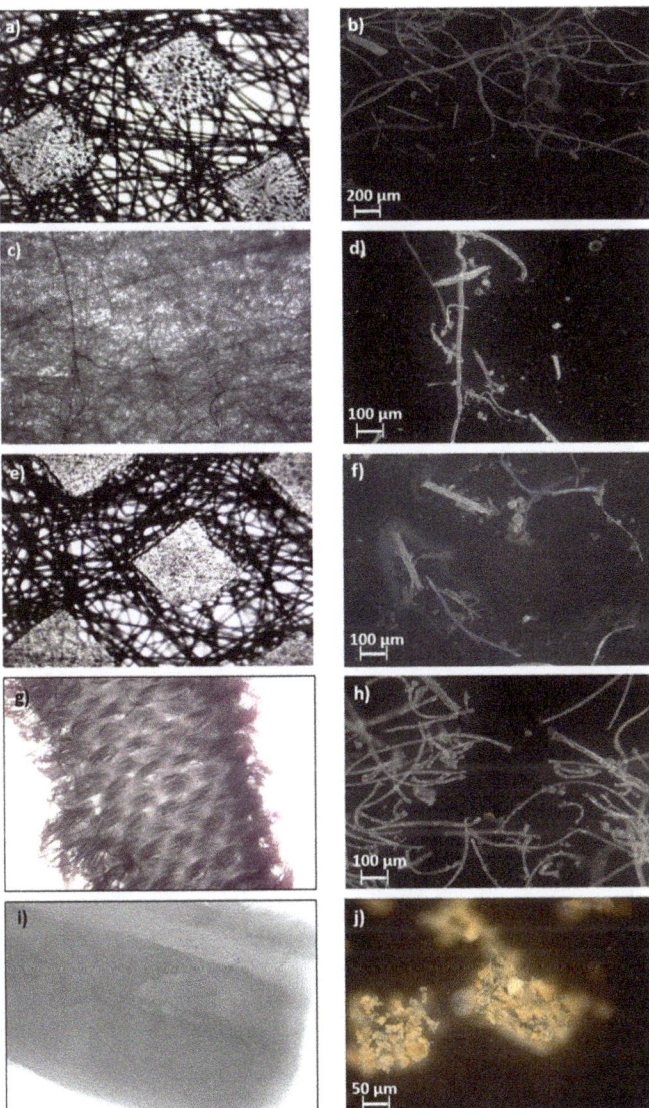

Figure 5. Optical microscope scans of pristine DSM parts before TCLP (magnified 4 times) and leached microplastic parts loaded on filters after TCLP (magnified 100 times): (**a**) pristine outer layer, (**b**) microplastic parts out of the outer layer, (**c**) pristine filter layer, (**d**) microplastic parts out of the filter layer, (**e**) pristine inner layer, (**f**) microplastic parts out of inner layer, (**g**) pristine ear loops, (**h**) microplastic parts out of ear loops, (**i**) pristine nose wire, (**j**) microplastic parts out of nose wire.

The TCLP products of the whole DSM sample were also analyzed by SEM (Figure 6a,b). Figure 6a show particles present in the analyzed leachate before filtration, and in Figure 6b they can be seen after filtration through filters with a mesh size of 0.2 µm. Accordingly, it is to be expected that in the first photo much larger particles can be seen, most of which are fibers that are uniformly distributed and superimposed. In Figure 6b, the particles are much smaller, different shapes, and much more evenly distributed, together with plastic particles in the nm range. In fact, when a DSM decomposes into nm range fragments (the latter are shown by SEM in Figure 6b), it is of particular environmental concern, as the release of nanoplastics is accelerated to due increased surface area.

Figure 6. Scans of micro- and nanoparticles released from whole DSM sample in TCLP process. (**a**) SEM image of particles loaded on the filter, (**b**) SEM image of filtrate, and (**c**) particle size distribution of filtrate.

In addition, to prove the leaching of nanoplastic particles in the filtrate of whole DSM leachate, particle size distribution (PSD) analysis, which was recorded three times, was performed (Figure 6c). With this analysis, the presence of nanoscale particles released by DSMs was confirmed. The particles have an average diameter of 305 nm. The majority of the particles (87%) have a diameter of 430 nm, with a standard deviation of 196.1 nm.

3.4. Photodegradability Results by Mimicking Natural Conditions with Xenotest

In general, disposable masks are exposed to the environment after use if not properly stored as waste, which means that the various environmental conditions and their influence should be studied over time. Therefore, in the following section, the DSMs were exposed to environmental conditions using a Xenotest climate chamber. After a long period of mimicking natural conditions (i.e., light, rain, etc., see Section 2.5), the DSMs were evaluated using various characterization techniques after mimicking natural weathering and compared to the untreated DSMs. It is very hard to assess the correlation between artificial weathering and natural weathering in general, as it depends on the type of material. According to Badji et al. [37], artificial weathering in a climate chamber is comparable to a 7.35 times longer period under natural weathering in equatorial regions. Further north and south from the equator, the acceleration time will be above this value. In accordance with Badji et al.'s study, it can be concluded that 200 cycles of artificial weathering, the duration of which was 100 h, are roughly equal to 735 h or 1 month under natural weathering in the equatorial region. Unlike other studies [18], which only investigated the change in properties of disposable masks under UV light weathering conditions, this study implements real daylight, rain, and other environmental parameters typical of real conditions.

The experiments on DSMs using Xenotest revealed that the changes are dramatic after artificial weathering. As is shown in Figure 7a, photodegradation of the three-layered part after 200 repetitions of a 30 min cycle is visually apparent (DSMs decay and become friable). This could be due to changes in chemical composition and fiber structure, which could be seen in the mask fragments broken into smaller pieces. The DSM polymer becomes more fragile, releasing even more microplastics and nanoplastics. In addition, it is suspected that the mechanical strength changed and most likely decreased. The test has shown that the three-layered part of the DSM loses more than 80% of its weight after 200 repetitions under simulated natural conditions in the Xenotest chamber as is shown in Figure 7b. Six samples of the three-layered part were analyzed (B–G) and, as can be seen from the graph of mass loss, the results of photodegradation are satisfyingly precise and show repeatability. This is also the part of the mask that degrades faster than other parts (Figure 7), as weight change in ear loops after 150 repetitions was 3.37% and after 200 repetitions it was 11.14%, and that of nose wires after 150 repetitions was 0.31% and after 200 repetitions it was 0.97%. These results show a significant influence of photodegradation of the masks, contributing to an increased number of micro/nanoplastics fragments in terms of weight change, confirmed by the DLS analysis below. Aged mask fragments have already been shown to completely transform into microplastics and it was predicted that a fully aged mask would release billions of microplastic fibers into the environment [17]. However, the contribution of more problematic nanoplastics has not been considered.

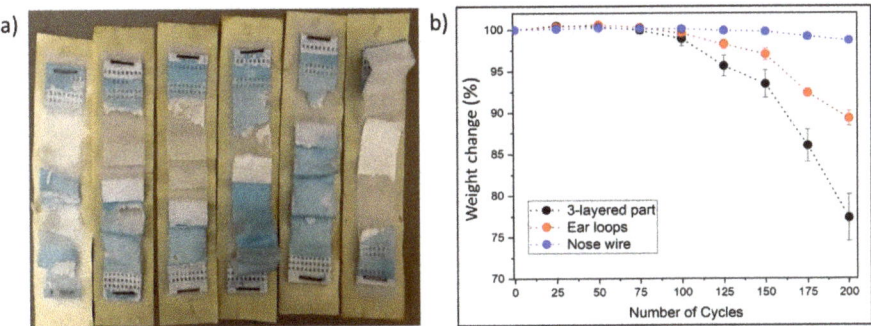

Figure 7. (a) Photodegradation analysis of DSM: (b) mass loss of DSM parts exposed to artificial weathering from 0 to 200 cycles.

FTIR analyses were performed before and after the Xenotest. The FTIR spectra changed only in the case of the three-layered part. In the case of the nose wire and ear loops, there were no visible changes in the collected spectra. Similar to the spectra collected from microplastics after the TCLP are the spectra of photodegraded DSMs (Figure 8). The FTIR spectrum of the photodegraded three-layered part does not agree with the spectrum of polypropylene and shows new peaks around 3300 cm^{-1}, 1730 cm^{-1}, and 1650 cm^{-1} and peaks in the range of 1030–1060 cm^{-1}. Listed signals at corresponding wavenumbers are typical for O-H, C=O, C-O, etc. vibrations [16,18]. The latter indicates that polypropylene undergoes oxidation/degradation reactions, as already observed above in Figure 4.

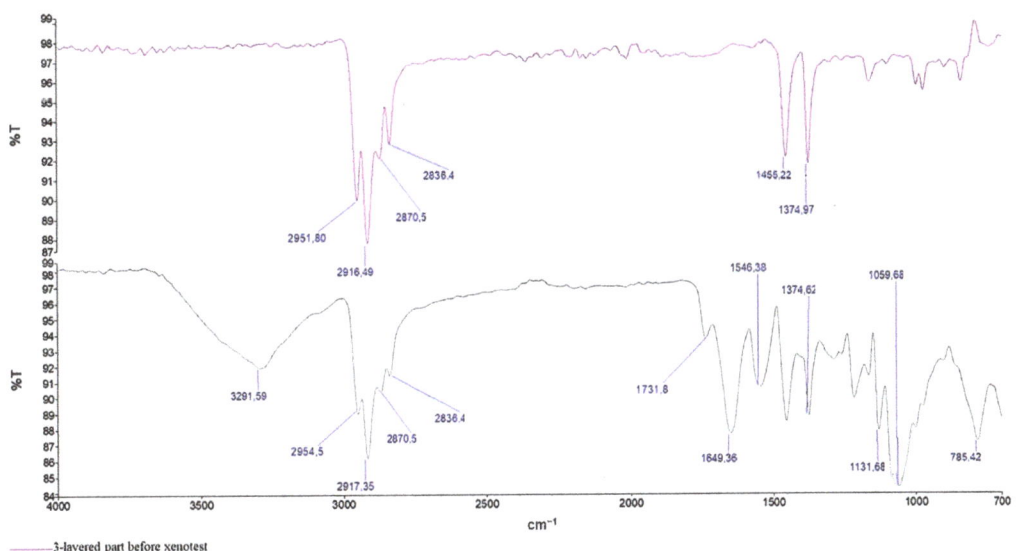

Figure 8. FTIR spectra of the 3-layered part before and after Xenotest.

The changes can also be seen in the TGA and DSC diagrams of the three-layer part of DSMs. The diagrams in Figure 9a,b show that 97% of the mass loss starts at lower temperatures—between 90 °C and 600 °C. The total mass loss from onset to 694.7 °C is 98.8%, in agreement with waste analysis (Table 2). Yet, after mimicking the real natural conditions in Xenotest, the three-layer part of DSMs after exposure seems less thermogravimetrically stable in comparison to virgin three-layer parts, as it loses its mass at a lower T than pristine three-layer parts. In addition, the approximate melting temperature of the DSM after the Xenotest is also lower—156 °C—as can be seen in Figure 9b. The phase transition flow T value for PP was determined to be around 160 °C [7], which coincides with our data and confirms that DSMs are mainly fabricated using PP polymer.

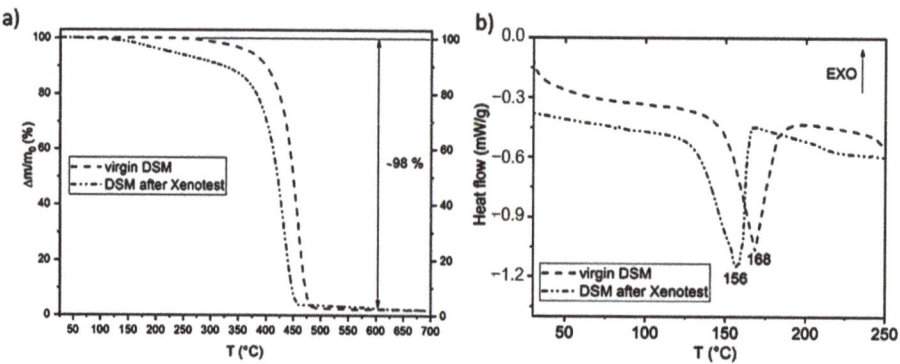

Figure 9. TGA (**a**) and DSC (**b**) analyses of whole DSM before and after Xenotest.

The simulated rainwater was collected during the test and analyzed under the optical microscope, which allowed us to isolate microplastic particles produced by photodegradation. As can be seen in Figure 10a, a DSM is degraded by photodegradation into extremely inhomogeneous microparticles, ranging from a few hundred micrometers to less than 1 µm. The formation of nanoplastics and their presence in the simulated rainwater were additionally confirmed by DLS analysis, the result of which can be seen in Figure 10b.

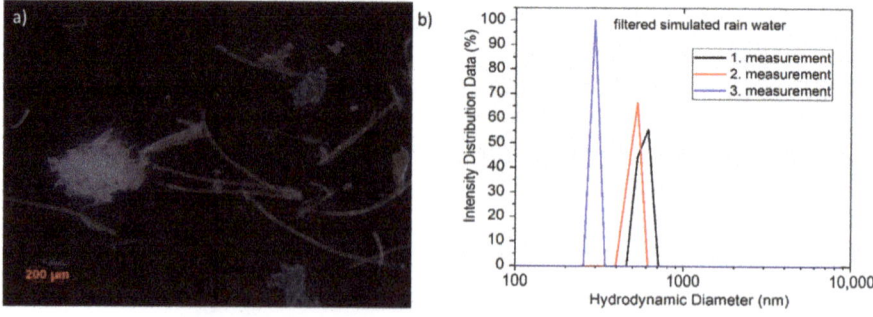

Figure 10. (**a**) Obtained microplastic parts from collected simulated rainwater; (**b**) DLS analysis of filtered simulated rainwater.

4. Discussion and Proposed Future Recommendations

To conclude, when we talk about DSM waste, we are talking about the waste that has been with us for a long time, but since the beginning of the epidemic, the quantities have increased dramatically. Given the volumes that are being generated, it is important to understand the wastes that we are dealing with. As we pointed out in the study, it is a homogeneous type of waste—each mask is made up of only five components and is primarily made of polypropylene. However, this should not lead us to forget about the awareness of the hazardous nature of the waste. As can be seen, this waste contains quite a high amount of Cl, which may indicate the presence of halogenated organic compounds (AOX). However, the main problem is the amount of microplastics and nanoplastics detected in this study, which are formed in a short time when this waste is exposed to natural conditions such as leaching, moisture, temperature, and UV radiation. Thus, it is highly recommended that these strategies are taken into consideration:

- Rapid fragmentation in the environment is an additional problem if this type of waste is not properly collected and processed. Given these considerations, proper disposal of DSM waste should be a priority for the system and the public [38]. In this way, the risks posed by the disposed masks could be better assessed and they could be prevented from entering the environment. Stricter management of mask waste would be beneficial, greatly limiting the amount of masks released into the environment. Such a setting would thus contribute to the control of micro/nanoplastics in the environment.
- A promising option that can open a new way to reduce the released fragments of micro/nanoplastics is recycling. Effective recycling could be another waste management option that would add value along with possible upgrades using various antiviral (nano)fillers. Another promising option would be to reuse the masks to minimize contamination. Reuse could be ensured by disinfection, but special care should be taken here to avoid damaging the fiber structures in the mask or reducing their efficiency in terms of protection or breathability.
- Additional research is needed to determine the best possible way to dispose of DSM waste, whether through energy or material recovery.
- Of particular importance would be bio-based and potentially biodegradable alternatives to current conventional plastic masks. Bio-based materials, including biopolymers, could partially or even completely replace petroleum-based polymers and in this way reduce the ecological footprint.
- Last, but not least, more consideration should be given to ecotoxicological measures. The need to assess the risk of released micro/nanoplastics in the environment should be investigated.

Author Contributions: A.E.: Formal analysis, Investigation, Methodology, Analysis, Visualization, Writing—original draft. O.P.: Analysis, Visualization, Data curation, Writing—review and editing. I.F.Z.: Funding acquisition, Supervision, Methodology, Writing—review and editing. J.V.V.: Conceptualization, Methodology, Funding acquisition, Formal analysis, Supervision, Writing—review and editing. All authors have read and agreed to the published version of the manuscript.

Funding: The authors would like to acknowledge the financial support provided by the Slovenian Research Agency (Grant Number: P2-0118 within COVID Programme and Young Researchers Programme and project No. J7-3149).

Institutional Review Board Statement: Not applicable.

Informed Consent Statement: Not applicable.

Data Availability Statement: Data available on request.

Conflicts of Interest: The authors declare that they have no known competing financial interest or personal relationships that could have appeared to influence the work reported in this paper.

References

1. Shashoua, Y. *Conservation of Plastics*; Taylor & Francis: London, UK, 2008. [CrossRef]
2. PlasticsEurope. *Plastics—the Facts 2019*; PlasticsEurop: Brussels, Belgium, 2019.
3. World Health Organization. *WHO Director-General's Opening Remarks at the Media Briefing on COVID-19-3 March 2020*; World Health Organization: Geneva, Switzerland, 2020.

4. Park, C.-Y.; Kim, K.; Roth, S.; Beck, S.; Kang, J.W.; Tayag, M.C.; Grifin, M. *Global Shortage of Personal Protective Equipment amid COVID-19: Supply Chains, Bottlenecks, and Policy Implications*; Asian Development Bank: Mandaluyong, Philippines, 2020. [CrossRef]
5. Czigany, T.; Ronkay, F. The coronavirus and plastics. *Express Polym. Lett.* **2020**, *14*, 510–511. [CrossRef]
6. Rathinamoorthy, R.; Balasaraswathi, S.R. Disposable tri-layer masks and microfiber pollution—An experimental analysis on dry and wet state emission. *Sci. Total Environ.* **2022**, *816*, 151562. [CrossRef] [PubMed]
7. Aragaw, T.A. Surgical face masks as a potential source for microplastic pollution in the COVID-19 scenario. *Mar. Pollut. Bull.* **2020**, *159*, 111517. [CrossRef] [PubMed]
8. Sun, J.; Yang, S.; Zhou, G.; Zhang, K.; Lu, Y.; Jin, Q.; Lam, P.; Leung, K.; He, Y. Release of microplastics from discarded surgical masks and their adverse impacts on the marine copepod Tigriopus japonicus. *Environ. Sci. Technol. Lett.* **2021**, *8*, 1065–1070. [CrossRef]
9. Spennemann, D.H.R. COVID-19 Face Masks as a Long-Term Source of Microplastics in Recycled Urban Green Waste. *Sustainability* **2022**, *14*, 207. [CrossRef]
10. Akber Abbasi, S.; Khalil, A.B.; Arslan, M. Extensive use of face masks during COVID-19 pandemic: (micro-)plastic pollution and potential health concerns in the Arabian Peninsula. *Saudi J. Biol. Sci.* **2020**, *27*, 3181–3186. [CrossRef] [PubMed]
11. Akhbarizadeh, R.; Dobaradaran, S.; Nabipour, I.; Tangestani, M.; Abedi, D.; Javanfekr, F.; Jeddi, F.; Zendehboodi, A. Abandoned Covid-19 personal protective equipment along the Bushehr shores, the Persian Gulf: An emerging source of secondary microplastics in coastlines. *Mar. Pollut. Bull.* **2021**, *168*, 112386. [CrossRef] [PubMed]
12. Allison, A.; Ambrose-Dempster, E.; Domenech, T.; Bawn, M.; Arredondo, M.; Chau, C.; Chandler, K.; Dobrijevic, D.; Hailes, H.; Lettieri, P.; et al. The environmental dangers of employing single-use face masks as part of a COVID-19 exit strategy. In *UCL Open: Environment Preprint*; UCL Press: London, UK, 2020. [CrossRef]
13. Ammendolia, J.; Saturno, J.; Brooks, A.L.; Jacobs, S.; Jambeck, J.R. An emerging source of plastic pollution: Environmental presence of plastic personal protective equipment (PPE) debris related to COVID-19 in a metropolitan city. *Environ. Pollut.* **2021**, *269*, 116160. [CrossRef] [PubMed]
14. Parashar, N.; Hait, S. Plastics in the time of COVID-19 pandemic: Protector or polluter? *Sci. Total Environ.* **2021**, *759*, 144274. [CrossRef] [PubMed]
15. Patrício Silva, A.L.; Prata, J.C.; Mouneyrac, C.; Barcelò, D.; Duarte, A.C.; Rocha-Santos, T. Risks of Covid-19 face masks to wildlife: Present and future research needs. *Sci. Total Environ.* **2021**, *792*, 148505. [CrossRef] [PubMed]
16. Saliu, F.; Veronelli, M.; Raguso, C.; Barana, D.; Galli, P.; Lasagni, M. The release process of microfibers: From surgical face masks into the marine environment. *Environ. Adv.* **2021**, *4*, 100042. [CrossRef]
17. Shen, M.; Zeng, Z.; Song, B.; Yi, H.; Hu, T.; Zhang, Y.; Zeng, G.; Xiao, R. Neglected microplastics pollution in global COVID-19: Disposable surgical masks. *Sci. Total Environ.* **2021**, *790*, 148130. [CrossRef] [PubMed]
18. Wang, Z.; An, C.; Chen, X.; Lee, K.; Zhang, B.; Feng, Q. Disposable masks release microplastics to the aqueous environment with exacerbation by natural weathering. *J. Hazard. Mater.* **2021**, *417*, 126036. [CrossRef] [PubMed]
19. Morgana, S.; Casentini, B.; Amalfitano, S. Uncovering the release of micro/nanoplastics from disposable face masks at times of COVID-19. *J. Hazard. Mater.* **2021**, *419*, 126507. [CrossRef] [PubMed]
20. SIST-TP CEN/TR 15310-4:2007; Characterization of Waste—Sampling of Waste Materials—Part 4: Guidance on Procedures for Sample Packaging, Storage, Preservation, Transport and Delivery. SIST: Newark, DE, USA, 2007.
21. Government of the Republic of Slovenia. *Rules on the Characterisation of Waste Prior to Landfill, the Characterisation of Hazardous Waste Prior to Incineration and the Performance of Chemical Analysis of Waste for Control Purposes*; 2016-01-2488; Government of the Republic of Slovenia: Ljubljana, Slovenia, 2016.
22. SIST EN 15002:2015; Characterization of Waste—Preparation of Test Portions from the Laboratory Sample. SIST: Newark, DE, USA, 2015.
23. SIST EN 15411:2011; Solid Recovered Fuels—Methods for the Determination of the Content of Trace Elements (As, Ba, Be, Cd, Co, Cr, Cu, Hg, Mo, Mn, Ni, Pb, Sb, Se, Tl, V and Zn). SIST: Newark, DE, USA, 2011.
24. SIST EN 13137:2002; Characterization of Waste—Determination of Total Organic Carbon (TOC) in Waste, Sludges and Sediments. SIST: Newark, DE, USA, 2002.
25. SIST EN 16168:2013; Sludge, Treated Biowaste and Soil—Determination of Total Nitrogen Using Dry Combustion Method. SIST: Newark, DE, USA, 2013.
26. Ebeling, M.E. The Dumas Method for Nitrogen in Feeds. *J. Assoc. Off. Anal. Chem.* **2020**, *51*, 766–770. [CrossRef]
27. SIST EN 15935:2012; Sludge, Treated Biowaste, Soil and Waste—Determination of Loss on Ignition. SIST: Newark, DE, USA, 2012.
28. SIST EN 15934:2012; Sludge, Treated Biowaste, Soil and Waste—Calculation of Dry Matter Fraction after Determination of Dry Residue or Water Content. SIST: Newark, DE, USA, 2012.
29. SIST-TS CEN/TS 16023:2014; Characterization of Waste—Determination of Gross Calorific Value and Calculation of Net Calorific Value. SIST: Newark, DE, USA, 2014.
30. SIST EN 15308:2017; Characterization of Waste—Determination of Selected Polychlorinated Biphenyls (PCB) in Solid Waste by Gas Chromatography with Electron Capture or Mass Spectrometric Detection. SIST: Newark, DE, USA, 2017.
31. SIST EN 14582:2017; Characterization of Waste—Halogen and Sulfur Content—Oxygen Combustion in Closed Systems and Determination Methods. SIST: Newark, DE, USA, 2017.
32. Method 1311 Toxicity Characteristic Leaching Procedure. EPA: Washington, DC, USA, 1992.

33. Potrč, S.; Kraševac Glaser, T.; Vesel, A.; Poklar Ulrih, N.; Fras Zemljič, L. Two-Layer Functional Coatings of Chitosan Particles with Embedded Catechin and Pomegranate Extracts for Potential Active Packaging. *Polymers (Basel)* **2020**, *12*, 1855. [CrossRef] [PubMed]
34. Glaser, T.K.; Plohl, O.; Vesel, A.; Ajdnik, U.; Ulrih, N.P.; Hrnčič, M.K.; Bren, U.; Fras Zemljič, L. Functionalization of Polyethylene (PE) and Polypropylene (PP) Material Using Chitosan Nanoparticles with Incorporated Resveratrol as Potential Active Packaging. *Materials* **2019**, *12*, 2118. [CrossRef] [PubMed]
35. Porubská, M.; Szöllős, O.; Kóňová, A.; Janigová, I.; Jašková, M.; Jomová, K.; Chodák, I. FTIR spectroscopy study of polyamide-6 irradiated by electron and proton beams. *Polym. Degrad. Stab.* **2012**, *97*, 523–531. [CrossRef]
36. Dincer, I.; Rosen, M.A.; Khalid, F. 3.16 Thermal Energy Production. In *Comprehensive Energy Systems*; Dincer, I., Ed.; Elsevier: Oxford, UK, 2018; pp. 673–706. [CrossRef]
37. Badji, C.; Soccalingame, L.; Garay, H.; Bergeret, A.; Bénézet, J.-C. Influence of weathering on visual and surface aspect of wood plastic composites: Correlation approach with mechanical properties and microstructure. *Polym. Degrad. Stab.* **2017**, *137*, 162–172. [CrossRef]
38. Remic, K.; Erjavec, A.; Volmajer Valh, J.; Šterman, S. Public Handling of Protective Masks from Use to Disposal and Recycling Options to New Products. *J. Mech. Eng.* **2022**, *68*, 281–289. [CrossRef]

Article

Water Environment Quality Evaluation and Pollutant Source Analysis in Tuojiang River Basin, China

Kai Zhang [1,*], Shunjie Wang [1], Shuyu Liu [1], Kunlun Liu [2], Jiayu Yan [1] and Xuejia Li [1]

1. School of Chemistry and Environment, China University of Mining and Technology (Beijing), Beijing 100083, China; sqt2000302077@student.cumtb.edu.cn (S.W.); bqt2100302035@student.cumtb.edu.cn (S.L.); 1910380329@student.cumtb.edu.cn (J.Y.); 1810380221@student.cumtb.edu.cn (X.L.)
2. Xinjiang Energy Co., Ltd. of State Energy Group, Wulumuqi 830000, China; 11390171@ceic.com
* Correspondence: zhangkai@cumtb.edu.cn; Tel.: +86-010-62339810

Abstract: A water environment quality evaluation and pollution source analysis can quantitatively examine the relationship among water pollution, resources, and the economy, and investigate the main factors affecting water quality. This paper took COD, NH_3-N, and TP of the Tuojiang River as the research objects. The water environment quality evaluation and pollution source analysis of the Tuojiang River Basin were conducted based on the grey water footprint, decoupling theoretical model, and correlation analysis method. The results showed that grey water footprint decreased, and the water environment quality improved. Among the pollution sources of the grey water footprint, TP accounted for the highest proportion. Moreover, the economic development level and the water environment were generally in a state of high-quality coordination. Farmland and stock breeding pollution accounted for the largest proportion of agricultural pollution and were thus the main source of the grey water footprint. The results of Pearson's correlation analysis indicated that the source of the pollutants were the imported pollution from the tributaries and agricultural pollution (especially stock breeding and farmland irrigation). These results showed that the quality of the water environment was improving, and the main factors affecting the water environment were stock breeding and farmland pollution in agriculture. This study presents a decision-making basis for strengthening the ecological barrier in the Yangtze River.

Keywords: water environment quality evaluation; pollution source analysis; grey water footprint; decoupling theoretical model; correlation analysis

Citation: Zhang, K.; Wang, S.; Liu, S.; Liu, K.; Yan, J.; Li, X. Water Environment Quality Evaluation and Pollutant Source Analysis in Tuojiang River Basin, China. *Sustainability* 2022, 14, 9219. https://doi.org/10.3390/su14159219

Academic Editor: Ning Yuan

Received: 1 July 2022
Accepted: 25 July 2022
Published: 27 July 2022

Publisher's Note: MDPI stays neutral with regard to jurisdictional claims in published maps and institutional affiliations.

Copyright: © 2022 by the authors. Licensee MDPI, Basel, Switzerland. This article is an open access article distributed under the terms and conditions of the Creative Commons Attribution (CC BY) license (https://creativecommons.org/licenses/by/4.0/).

1. Introduction

Given that the Yangtze River is the largest river in China, the construction and development of the Yangtze River Basin plays an important role in the national sustainable development strategy [1,2]. Most areas of the Yangtze River discharge giant amounts of sewage, and environmental issues are becoming additionally apparent, particularly pollution and the deterioration of water quality [3]. As an important first-order tributary on the right bank of the upper reaches of the Yangtze River, the Tuojiang River is a vital ecological barrier in the upper reaches of the Yangtze River Basin and plays a highly critical role in maintaining the ecological security of the Yangtze River Economic Belt [4–6]. Therefore, the evaluation of the water environment quality and analysis of the pollution sources of cities along the Tuojiang River Basin are imperative, and timely and targeted measures ought to be taken for prevention. Such studies can provide a decision-making basis for strengthening the ecological barrier in the upper reaches of the Yangtze River.

Traditional water environment quality evaluation methods mainly include the pollution index evaluation method [7], single factor evaluation method [8], and artificial neural network analysis method [9,10]. However, these approaches mainly evaluate the pollution degree of polluted water bodies, and studies on the relationship between the quantity and quality of water resources are few. In recent years, a comprehensive analysis of water resources that mixes water quality and quantity has emerged. Tharme [11] used the science of environmental flow assessment to determine the quantity and quality of water needed for ecosystem conservation and resource conservation. Xia et al. [12] established a comprehensive evaluation method for water quantity and quality in a basin and proposed the integration concept of water resource functional capacity and water resource functional deficit. Wang et al. [13] took the Liaohe River as an example to establish a comprehensive evaluation method for ecological water demand that considers the natural and social water cycles, as well as the river water quantity and quality. However, the above methods still fail to quantitatively explain the effect of water pollution on the quantity of water resources. The concept of grey water footprint (GWF) provides a new idea for the quantitative evaluation of the relationship between water quantity and quality, and grey water footprint was used to measure water pollution levels [14]. Grey water footprint was defined as the amount of freshwater necessary for the pollutant load to be assimilated to reach the level of existing water quality standards [15,16]. Grey water footprint has been used to assess the influences of global human economic activities on water use and has been widely developed to assess water pollution levels in many fields, such as agricultural grey water footprint (GWF_{Agr}) [16–20], industrial grey water footprint (GWF_{Ind}) [21–23], and domestic grey water footprint (GWF_{Dom}) [24]. Previous research directions have focused on the grey water footprint of specific pollutants (e.g., nitrogen-related or phosphorus-related grey water footprint) for developing control strategies [25–29]. Only a few studies have considered multiple pollutants to show the overall picture of pollution [30]. Previous analyses of the grey water footprint for multiple pollutants have focused on pollution levels, whereas studies on the dynamic changes of major water pollutants that lead to water pollution are rare [31]. In this paper, the grey water footprint of various pollutants in the Tuojiang River Basin (Neijiang City (NJC) section) was studied, and the sources of grey water footprint in agriculture, industry, and domestic areas were analyzed. The dynamic changes of the grey water footprint of various pollutants and their sources were also investigated. Some studies have introduced decoupling theoretical models based on investigating the grey water footprint to reveal the decoupling relationship between economic development and the water environment [32]. These studies have comprehensively evaluated the quality of the water environment. The present work studied the grey water footprint of various pollutants in the Tuojiang River Basin (NJC section) and analyzed the sources of grey water footprint in agricultural, industrial, and domestic areas. In addition, the dynamic changes of the grey water footprint of multiple pollutants and their sources were studied, and the water environment quality was comprehensively evaluated by combining grey water footprint and decoupling theory.

Identifying pollution sources and determining their contribution are the basis for the effective prevention of pollution [33–35]. Grey water footprint can show the overall level of water pollution but cannot trace the specific source of pollutants in the river. Many scholars have investigated the sources of pollutants using the correlation analysis method for qualitative source analysis [36–38]. For example, a correlation analysis in the Yangtze River Basin showed that the concentration of antibiotics in surface water and sediment is strongly correlated with total nitrogen and total phosphorus, which may have come from household and agricultural waste [39]. Zhang et al. [40] used a correlation analysis method to prove that the pollution sources of sulfate and fluoride in groundwater in a certain area in southwest China are highly similar and are greatly affected by the discharge of industrial parks. The source apportionment of pollutants was performed based on the correlation degree of the main pollutants in each section and the correlation degree of the main pollutants in the river with indicators of agriculture, forestry, animal husbandry, fishery, industry, and population economy.

This paper further evaluated the water environment quality and analyzed the pollution sources based on the grey water footprint, decoupling theoretical model, and correlation analysis method of the Tuojiang River. It also determined the status of the water environment quality and the main factors affecting the water environment. In addition, it provided a clear direction for reducing the grey water footprint and improving water quality. The main objectives of this study are to: (i) analyze the overall change in water environment quality and its decoupling from economic development, (ii) examine the main pollutants that affect the quality of the water environment, (iii) investigate the specific sources of water environmental quality, and (iv) provide recommendations for reducing the grey water footprint and improving the water quality of the Tuojiang River.

2. Materials and Methods

2.1. Study Areas and Data Sources

As shown in Figure 1, the Tuojiang River is a first-order tributary of the upper reaches of the Yangtze River and one of the more important rivers in the central area of Sichuan Province. NJC is located in the southeastern part of Sichuan Province and the middle part of the lower reaches of the Tuojiang River. The Tuojiang River Basin accounts for more than 95% of the city's land area. NJC includes Shizhong District (SZ), Dongxing District, Weiyuan County, Zizhong County (ZZ), and Longchang County (LC) [41]. The results of the 2021 Neijiang National Economic and Social Development Statistical Bulletin indicated that the economy of NJC has developed rapidly. In 2021, the gross domestic product (GDP) of NJC increased by 8.5% compared to the previous year. Specifically, the primary, secondary, and tertiary industries increased by 6.9%, 6.5%, and 10.4%, respectively, with a ratio of close to 17:33:60. This finding showed that the service industry makes up most of NJC's economy, whereas agriculture makes up the least.

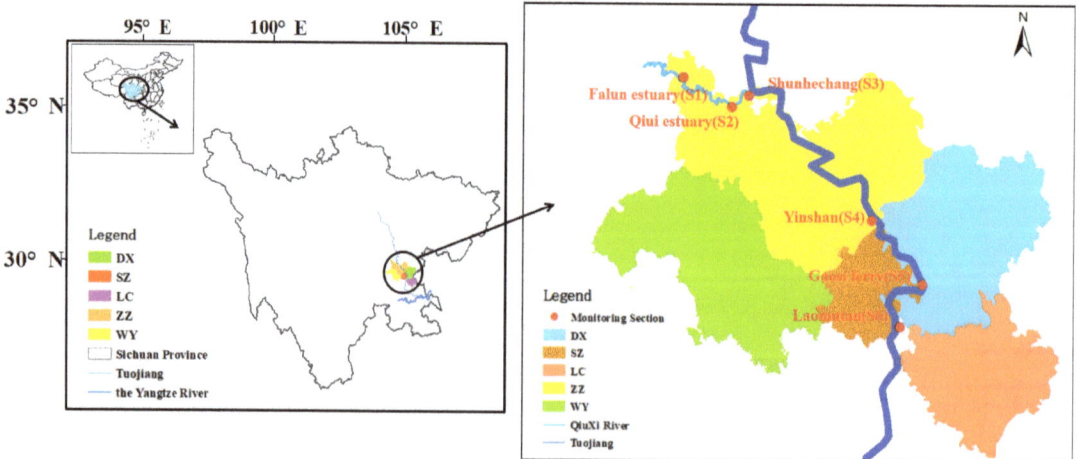

Figure 1. Geographical location of study areas.

The Tuojiang River enters the Neijiang River from Shunhechang and exits at Laomutan. The larger tributaries in the NJC section of the mainstream of the Tuojiang River mainly includes the Qiuxi, Mengxi, Xiaoqinglong, and Daqingliu Rivers. In the mainstream control unit of the Tuojiang River, the Qiuxi River is the first tributary of the Tuojiang River in the Neijiang River. The spatial and temporal distributions of the water resources in the Tuojiang River (NJC section) is uneven, and most of the precipitation is from June to September each year. In this paper, the monthly pollutant monitoring data of typical sections were selected to analyze the distribution characteristics of the main pollutants in the river (Table 1).

Table 1. List of typical monitoring sections.

Monitoring Section Name	Number	Section Properties
Falun estuary	S1	Entry section
Qiuxi estuary	S2	Section of the estuary entering the Tuojiang River
Shunhechang	S3	Entry section
Yinshan Town	S4	control section
Gaosi ferry	S5	control section
Laomutan	S6	Exit section

The agricultural, industrial, and domestic pollution data of each district and county in NJC are all from the Neijiang Statistical Yearbook (2015–2021). The data of the main pollutants in each section of the river came from the Neijiang Municipal Bureau of Ecology and Environment.

2.2. Major Pollutant Identification

In this study, the chemical oxygen demand (COD), ammonia nitrogen (NH_3-N), and total phosphorus (TP) were selected to evaluate the grey water footprint and analyze the changes in the major pollutants in the Tuojiang River for three main reasons: (1) According to the China Ecological Environment State Bulletin, COD, NH_3-N, and TP are the main pollutants causing surface water pollution in China; (2) according to the Sichuan Province news and documents, the main pollutants of rivers are COD, NH_3-N, and TP; and (3) previous studies on the Tuojiang River have shown that the main pollutants include COD, NH_3-N, and TP [42,43].

2.3. Grey Water Footprint

Grey water footprint can be interpreted as the dilution water demand [26]. When multiple pollutants are involved, the grey water footprint takes the maximum value of the grey water footprint of each pollutant [18,23,44]:

$$GWF_{j,i} = L_{j,i}/(C_{max,j,i} - C_{nat,j,i}), \tag{1}$$

$$GWF_j = \max\{GWF_{j,1}, GWF_{j,2}, \ldots, GWF_{j,i}\}, \tag{2}$$

$$GWF_{Total} = \Sigma GWF_j, \tag{3}$$

where $GWF_{j,i}$ is the grey water footprint of pollutant i released into the water at point j [volume/time], GWF_j is the grey water footprint of the pollutant at point j [volume/time], GWF_{Total} is the grey water footprint of the system being studied [volume/time], $L_{j,i}$ is the quantity of pollutant i being emitted into the water at point j [weight/volume], $C_{max,j,i}$ is the maximum permissible concentration of substance i in the receiving waters at point j [weight/time], $C_{nat,j,i}$ is the natural concentration of substance i in the receiving waters at point j [weight/volume], and n is the number of discharge points.

The natural water body background concentration (C_{nat}) of pollutants is the original concentration of pollutants in the water body, which is often assumed to be 0.

2.4. Water Environment Decoupling Theory

Decoupling theory refers to a situation in which the relationship between two or more related variables decreases or ceases to exist [45,46]. The decoupling effect refers to reducing the burden on the environment while maintaining or increasing economic growth rates. This effect enables more efficient use of natural resources, especially water [47]. Decoupling can be divided into relative and absolute decoupling. The former refers to the increase of resource and environmental pressure at a lower rate during economic growth; the latter refers to the reduction of the growth rate of resource and environmental pressure during economic growth [48]. According to the decoupling index, the gross domestic product change rate, and the grey water footprint change rate, the relationship between economic development and the water environment is divided into high-quality coordination, preliminary coordination, and uncoordinated [49]. The decoupling index can be calculated as follows:

$$DF = VGDP - VF, \qquad (4)$$

where DF is the decoupling index, VGDP is the average annual gross domestic product change rate, and VF is the average annual grey water footprint change rate. When VGDP > 0, VF < 0, and DF > 0, water resource utilization and economic growth are in a state of strong decoupling and high-quality coordinated development. When VGDP > 0, VF > 0, and DF > 0, water resource utilization and economic growth are in a weak decoupling and preliminary coordination state. When DF $\leq$ 0, water resource utilization and economic growth are not in a state of decoupling.

2.5. Correlation Analysis Method

A correlation analysis can indicate whether the variables are correlated, the direction of correlation, and the degree of closeness. Probability (P) reflects the probability of an event happening. Generally, $p < 0.05$ indicates a significant correlation; whereas $p < 0.01$ denotes an extremely significant correlation, which implies that the probability that the difference between samples is caused by a sampling error of less than 0.05 or 0.01. Pearson's correlation coefficient (Cor) was used to measure the correlation between two variables, X and Y. Its value is between -1 and 1, and the greater the absolute value is, the greater the correlation will be. A positive correlation coefficient indicates a positive correlation; otherwise, it is a negative correlation [50,51]. The formula for calculating Cor is [52]:

$$\mathrm{Cor}(X, Y) = \mathrm{Cov}(X, Y) / \sqrt{\mathrm{Var}(X)\mathrm{Var}(Y)}, \qquad (5)$$

where Cov(X, Y) denotes the co-variance of X and Y for any random variable Z, and Var (Z) denotes the variance of Z. On the basis of Cauchy–Schwarz inequality, Cor (X, Y) $\in [-1, 1]$.

3. Results and Discussion

3.1. Water Environment Quality Evaluation

3.1.1. Distribution and Proportion of Grey Water Footprint

The GWF_{Total}, GWF_{COD}, GWF_{NH3-N}, and GWF_{TP} of NJC and its five districts and counties are shown in Figure 2. From 2015 to 2019, the grey water footprint data of pollutants showed that GWF_{TP} was higher than GWF_{COD} and GWF_{NH3-N}; thus, GWF_{TP} was the grey water footprint, that is, the main pollutant was TP. The GWF_{Total} (Figure 2a) of NJC decreased from 556.82×10^8 m^3/y to 428.11×10^8 m^3/y, showing a decrease rate of 30.06%—and the GWF_{COD} (Figure 2b), GWF_{NH3-N} (Figure 2c), and GWF_{TP} (Figure 2d) decreased by 27.67%, 15.70%, and 42.04%, respectively. The grey water footprint of the five districts and counties generally showed a downward trend. Generally, the quality of the water environment is improving. The largest declines were in SZ and LC, which were 96.18% and 51.8%, respectively. The smallest decline was in ZZ at 6.63%. From the perspective of spatial distribution, the grey water footprint was significantly higher in ZZ and significantly lower in SZ. Thus, the water environment quality in the upper reaches of the Tuojiang River was slightly lower than that in the lower reaches. The phosphorus industry was a major factor leading to excessive pollutant concentrations in the environment. Several mining regions can be found in the upper reaches of the Tuojiang River, and their effects on water quality shouldn't be underestimated. The development of the phosphorus industry in the upper reaches of the Tuojiang River could explain the high TP concentration throughout the whole basin [43].

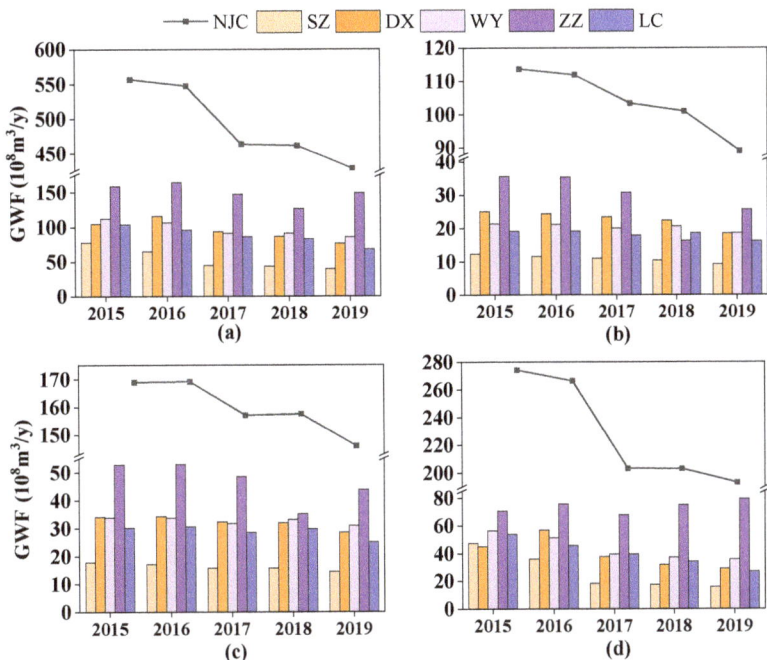

Figure 2. GWF_{Total} (**a**), GWF_{COD} (**b**), GWF_{NH_3-N} (**c**) and GWF_{TP} (**d**) of cities and counties.

The changes in the pollution sources of GWF_{COD}, GWF_{NH_3-N}, and GWF_{TP} in NJC from 2015 to 2019 are shown in Figure 3. The results showed that in comparison with 2015, the GWF_{Agr} in GWF_{COD} in 2019 decreased by 26.34×10^8 m^3, GWF_{Ind} decreased by 1.84×10^8 m^3, and GWF_{Dom} increased by 3.54×10^8 m^3; thus, the main reason for the decrease in GWF_{COD} was agricultural COD emissions (Figure 3a). The GWF_{Agr} in GWF_{NH_3-N} decreased by 23.78×10^8 m^3, GWF_{Ind} decreased by 2.21×10^8 m^3, and GWF_{Dom} increased by 3.07×10^8 m^3; thus, the main reason for the decrease in GWF_{NH_3-N} was agricultural NH$_3$-N emissions (Figure 3b). In GWF_{TP}, GWF_{Agr} decreased by 9.16×10^8 m^3, GWF_{Ind} decreased by 72.2×10^8 m^3, and GWF_{Dom} was almost unchanged as a whole. Hence, the main reason for the decline in GWF_{TP} was the reduction in industrial TP emissions (Figure 3c).

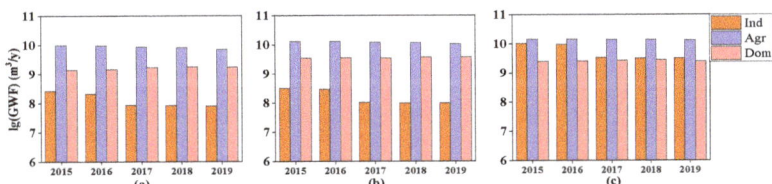

Figure 3. Pollution sources of GWF_{COD} (**a**), GWF_{NH_3-N} (**b**), and GWF_{TP} (**c**).

The main reason for the decline of GWF_{Ind} was that in 2016, NJC was the opportunity for Neijiang to seize the "One Belt, One Road" strategy and vigorously implement the innovation-driven strategy, which achieved remarkable results. Subsequently, NJC was committed to the transformation and upgrade of traditional industries, which entailed adjusting the industrial structure, actively promoting the development of new industrialization, and accelerating the cultivation of high-end growth industries, such as energy conservation, environmental protection, and shale gas. Furthermore, it required adherence to the development of new materials, new equipment, new energy, and other related industries as the main direction of economic development.

The proportion of pollutants from the GWF_{Total} in NJC from 2015 to 2019 are shown in Figure 4a. The proportion of the main sources of GWF_{COD}, GWF_{NH3-N}, and GWF_{TP} is shown in Figure 4b–d, respectively. As shown in Figure 4a, from 2015 to 2019, significant differences were observed in the proportion of pollutants in the GWF_{Total} in NJC. The average proportions of GWF_{COD}, GWF_{NH3-N}, and GWF_{TP} were 21%, 33%, and 46%, respectively. The proportion was $GWF_{TP} > GWF_{NH3-N} > GWF_{COD}$; thus, GWF_{TP} was the grey water footprint of NJC. The main sources of GWF_{COD}, GWF_{NH3-N}, and GWF_{TP} were agricultural pollution, with an average proportion of 83%, 76%, and 63%, respectively (Figure 4b–d). The second source of GWF_{COD} and GWF_{NH3-N} was domestic pollution. However, the second source of GWF_{TP} was mainly industrial pollution in 2015–2016, and the proportion of domestic and industrial pollution was approximately equal during 2017–2019. In recent years, relevant studies have been conducted on the driving force analysis of grey water footprint at the provincial scale in China, and agricultural activities have the highest contribution rate to the country's grey water footprint [53,54]. Given the continuous improvement in the level of agricultural development, organic or inorganic pollutants, such as nitrogen, phosphorus, and pesticides, enter the surface water, groundwater, and soil environment through surface runoff.

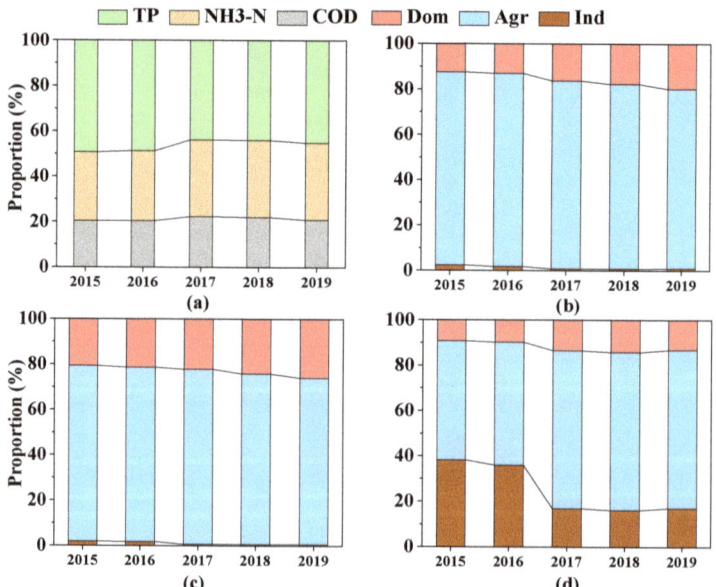

Figure 4. The proportion of pollution sources of GWF_{Total} (a), GWF_{COD} (b), GWF_{NH3-N} (c), and GWF_{TP} (d).

The changes in the specific sources of agricultural pollution are shown in Figure 5. We found that the source of agricultural pollution in the Tuojiang River Basin mainly came from stock breeding and farmland pollution, which was consistent with the research results of Hu Yunyun [55]. The extensive agricultural production methods and the discharge of livestock and poultry breeding wastewater led to the development of production-type pollution in the Tuojiang River Basin, that is, stock breeding and farmland pollution [56]. From 2015 to 2019, the proportion of agricultural pollution in GWF_{COD} and GWF_{NH3-N} decreased due to the reduction in livestock production (Figure 5a,b). According to the first national pollution source census, aquaculture accounted for a large proportion of agricultural water pollution. The grey water footprint source in aquaculture was wastewater pollution caused by animal urine and feces. COD and N were usually pollutants discharged from this wastewater [53]. Although the contribution of aquaculture was extremely low, the accumulation of discharged nutrients in aquatic systems could have a negative effect on water quality [57]. The deterioration of eutrophication caused by aquaculture and the resulting "red tide" problem could not be ignored [57,58]. In the agricultural production activities in the Tuojiang River Basin, excessive chemical fertilizers and irrigation cause agricultural planting pollution, and stock breeding produces a high pollution load. In recent years, the livestock and poultry breeding industry and the aquaculture industry have developed rapidly in townships and villages, and the wastewater from the aquaculture industry has surged. Therefore, to reduce the effect on river water quality, attention should be paid to the problem of direct discharge of wastewater from farms without treatment or if the treatment does not meet standards [59].

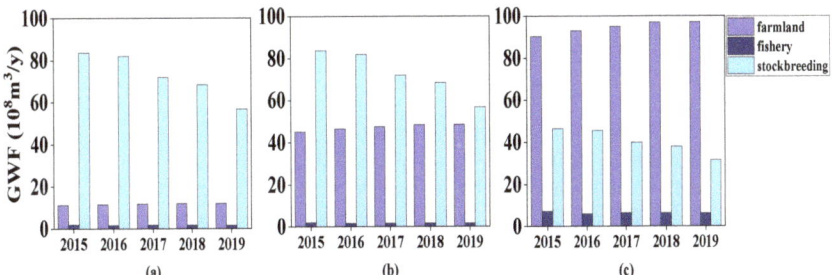

Figure 5. Specific sources of agricultural pollution for GWF_{COD} (**a**), GWF_{NH3-N} (**b**), and GWF_{TP} (**c**).

The proportion of agricultural pollution in the GWF_{TP} increased due to the rise in the irrigated area of farmland (Figure 5c). Some studies have shown that the pollution load of agricultural irrigation return water is mainly phosphorus, and the pollutants from farmland irrigation return water enter small river channels and ponds in villages through field ditches, and finally enter rivers [60,61]. A large amount of cultivated land exists on both sides of the mainstream of the Tuojiang River. Rainfall and farmland irrigation flushes the chemical fertilizer residues in the farmland into the Tuojiang River with surface runoff, which directly affects the water quality of the Tuojiang River. Agricultural non-point source pollution should be controlled, the abuse of chemical fertilizers and pesticides in agricultural production should be avoided, and efficient fertilization and irrigation technology should be promoted.

3.1.2. Decoupling of Economy and Water Environment

The decoupling of water environment quality and economic development in the Tuojiang River is shown in Figure 6. From 2016 to 2019, the decoupling index and the gross domestic product rate of change were positive, and the grey water footprint was negative. The order of the decoupling index was 2017 > 2019 > 2018 > 2016, with decoupling indices of 0.096, 0.325, 0.117, and 0.120, respectively. In recent years, the water resource environment and economic development have been in a state of absolute decoupling. The economic development level and water environment of the Tuojiang River were generally in a state of high-quality coordination. This phenomenon indicates that the economic development level was relatively high and the damage to the water environment was relatively low. The current economic development was in a growth trend, but the grey water footprint was gradually declining. The main reason for economic development was that the 13th Five-year Plan of the South Sichuan Economic Zone clearly proposed to accelerate the development of NJC. The development plan started to focus on the improvement of equipment manufacturing, advanced materials, electronic information, energy and chemical industries, and food and beverage industries, and was coordinated to build an important metropolitan area on the main axis of Chongqing's development.

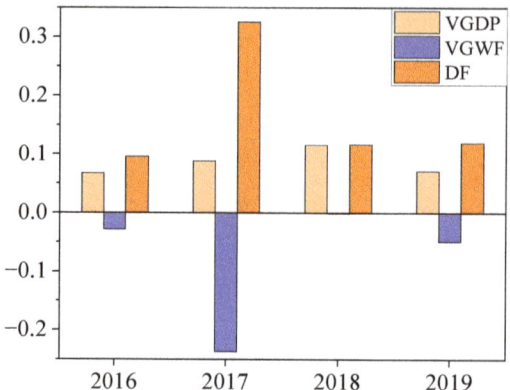

Figure 6. The decoupling of water environment quality and economic development.

The economic development and water environment of the Tuojiang River Basin were in a state of high-quality coordination. Thanks to the transformation and upgrading of coal energy development, it had embarked on the road of green and low-carbon development. NJC had a solid foundation for the development of green and low-carbon industries. Shale gas, hydrogen energy, and other clean energy sources were abundant, and the advantages of resource endowment and industrial bases made it easy to form a good trend of industrial green development. In the context of industrial transformation, economic development has been improved and pollution emissions have decreased.

3.2. Pollution Source Analysis of River

3.2.1. Distribution Characteristics of Pollutants in Sections

The seven-year average over-standard rate of pollutants is shown in Figure 7a. Exceeding the Class III water standard in the Environmental Quality Standard for Surface Water (GB3838-2002) is considered to exceed the standard—that is, COD > 20 mg/L, NH_3-N > 1.0 mg/L, TP > 0.2 mg/L. The excess rate is the ratio of the number of months exceeded to the total number of months. The COD exceeding rates of S1–S6 were 7.14%, 6.33%, 2.38%, 0%, 2.38%, and 0%, respectively, and were ranked as S4 = S6 < S5 = S3 < S2 < S1. The NH_3-N exceeding rates of S1–S6 were 15.48%, 3.57%, 1.19%, 0%, 0%, and 1.19%, respectively, and were ranked as S4 = S5 < S3 = S6 < S2 < S1. The TP exceeding rates of S1–S6 were 71.43%, 48.10%, 48.81%, 22.62%, 25%, and 23.81%, respectively, and were ranked as S4 < S6 < S5 < S2 < S3 < S1. Particularly, the COD of S4 and S6 and the NH_3-N of S4 and S5 did not exceed the standard. Therefore, the quality of COD and NH_3-N of S3–S6 was better, and the quality of COD and NH_3-N of S1 and S2 was poor. From 2014 to 2020, the average value of COD, NH_3-N, and TP of S1–S6 was 15.55 mg/L, 0.31 mg/L, and 0.23 mg/L, respectively. Therefore, the TP in the sections were more serious [5]. After the investigation, the reasons for the serious TP exceeding the standard were found as follows. The TP in the entry water quality was seriously exceeding the standard. A large amount of phosphogypsum was stored irregularly in some areas, and the source of phosphogypsum released a large amount of TP, which seriously polluted the water environment [42,62]. The phosphorus removal technology of urban sewage treatment facilities needed to be transformed and upgraded. After the transformation, the amount of TP pollutants entering the river could be further controlled and reduced.

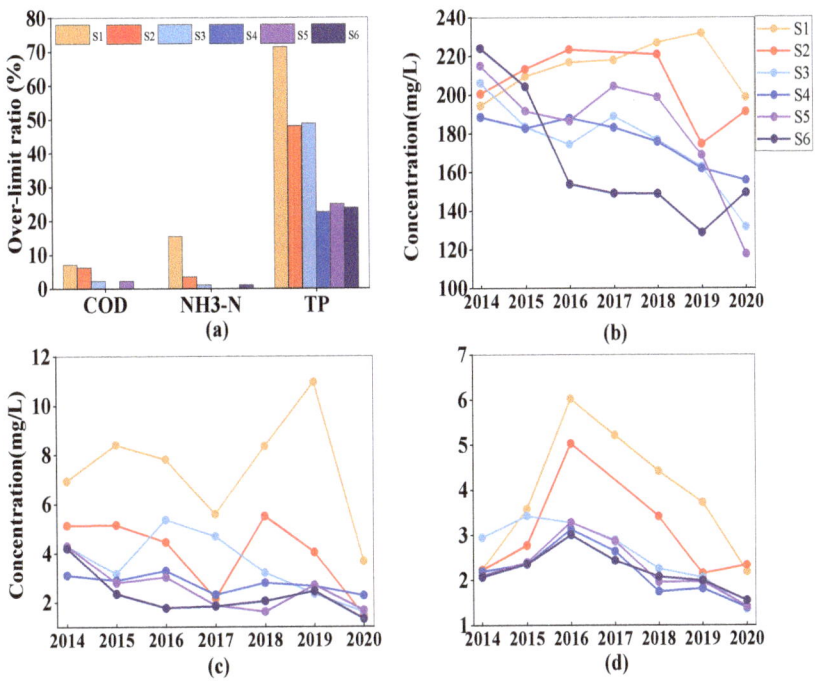

Figure 7. Average over-standard rate (**a**), annual changes of COD (**b**), NH$_3$-N (**c**), and TP (**d**) in each section.

The annual changes of pollutants in each section are shown in Figure 7b–d. From 2014 to 2020, the average annual total COD values of S1–S6 were 213.91, 204.18, 175.15, 176.74, 183.46, and 165.60 mg/L, respectively (Figure 7b). The average annual total NH$_3$-N values were 7.39, 3.99, 3.53, 2.77, 2.58, and 2.29 mg/L, respectively (Figure 7c). The annual mean values of TP were 3.91, 2.99, 2.61, 2.19, 2.29, and 2.21 mg/L, respectively (Figure 7d). The annual total values of COD, NH$_3$-N, and TP of S1 and S2 were obviously higher in the six sections, and the overall trend was roughly the same, which belonged to imported pollution. From the overall analysis, the annual total value of NH$_3$-N of S1 and S2 was at a turning point in 2017, and the annual total value of TP was at a turning point in 2016. The total amount of major pollutants in S3–S6 generally increased initially and then decreased.

The water-period changes of pollutants in each section are shown in Figure 8. The average values of the pollutant concentrations of the six sections indicated that the COD, NH$_3$-N, and TP concentrations were 16.34, 0.59, and 0.25 mg/L in a typical month of the dry season in March; 14.29, 0.23, and 0.19 mg/L in a typical month of the mid-dry season in November; and 15.16, 0.18, and 0.20 mg/L in a typical month of the wet season in August, respectively. Therefore, the average value of each pollutant in a typical month in the dry season was the highest, and the wet season was similar to the mid-dry season, which was lower than that in the dry season. This finding could be attributed to the increased precipitation within the wet season and the sturdy dilution of rain. In addition, the water temperature was high during the wet season, and the proliferation of microorganisms would degrade pollutants [63].

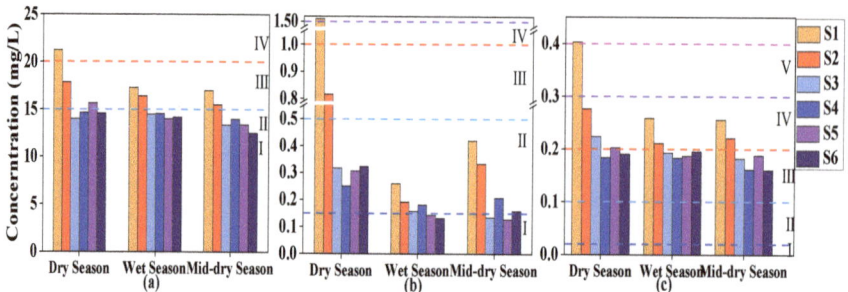

Figure 8. The water-period changes of COD (**a**), NH$_3$-N (**b**), and TP (**c**) in each section (I–V represented the Class I–V water standard in the Environmental Quality Standard for Surface Water (GB3838-2002)).

3.2.2. Relationship of Pollutants between Sections

The correlation analysis of pollutants in each section is shown in Figure 9. The NH$_3$-N between the six sections showed significant or extremely significantly positive correlations (Figure 9a). Particularly, the positive correlation between S1 and S2 was the most significant. From the perspective of the flow direction of the river and the spatial distribution of the cross sections, S1 was the only cross-sectional upstream of S2 and was unaffected by other tributary cross sections. Given that the NH$_3$-N of S1 was higher, the correlation degree between the Qiuxi and Falun estuaries was the highest, which also showed that the NH$_3$-N pollution of S2 came from the imported pollution of S1. Extremely significant positive correlations existed between S4 and S3, S5 and S4, and S6 and S5. The correlation between the downstream section and the most adjacent upstream section was high, which indicated that the NH$_3$-N downstream of the mainstream of the Tuojiang River (NJC section) was more susceptible to the upstream influence. In conclusion, the NH$_3$-N pollution in the downstream of the Qiuxi River (NJC section) might have originated from the imported pollution upstream. The NH$_3$-N pollution in the downstream of the mainstream of the Tuojiang River (NJC section) was easily affected by the upstream of the mainstream of the Qiuxi River.

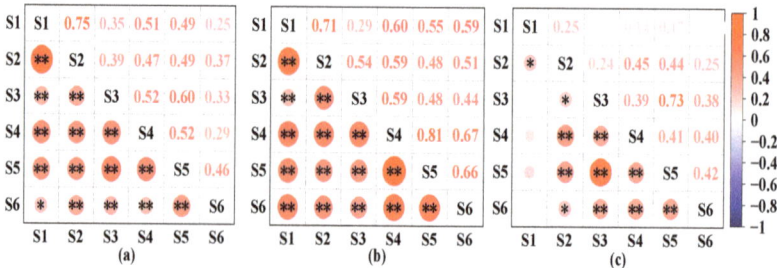

Figure 9. The correlation analysis of NH$_3$-N (**a**), TP (**b**), and COD (**c**) (* $p \leq 0.05$, ** $p \leq 0.01$).

A highly significantly positive correlation existed between the TP monitoring data of the six sections, and the correlation degree was generally higher than those of NH$_3$-N and COD (Figure 9b). Particularly, the correlation degree of TP between S4 and S5 was the highest, and that between S1 and S2 was also higher. Thus, the TP pollution in the lower reaches of the Qiuxi River (NJC section) possibly originated from the imported pollution upstream, and the water quality of the mainstream of the Tuojiang River (NJC section) was greatly affected by the adjacent sections.

A significantly positive correlation of COD existed between S1 and S2 (Figure 9c). This result was consistent with the correlation analysis of NH$_3$-N and TP, indicating that the COD pollution of S2 came from the input pollution of S1. S1 was significantly correlated with S2, but not significantly correlated with the COD of other sections; thus, the downstream section was mostly affected by the adjacent upstream section. A significant positive correlation existed among S2–S6, which indicated that the COD pollution downstream of the mainstream of the Tuojiang River (NJC section) was easily affected by the upstream water quality. Given the long distance between S2 and S6, the correlation strength was relatively weak. S3 is located at the intersection of the Qiuxi River and the mainstream of the Tuojiang River, and the water quality obviously fluctuates. Although S3 was similar to S2 in spatial distance, the correlation between S3 and S2 was not strong.

3.2.3. Correlation Analysis between Pollutants and Industry Indicators

The correlation analysis between pollutants and industry indicators is shown in Figure 10. As shown in Figure 10a, at the $p < 0.05$ level, NH_3-N was positively correlated with indicators for agriculture and forestry. The range of the correlation coefficient was 0.827–0.881, and the order of the correlation degree from small to large was park, woodland, and grassland area (0.827) < cultivated area (0.828) < fertilizer application (0.856) < farm irrigation (0.874) < crop yield (0.881). Thus, the contribution of each pollution source to NH_3-N emission was in the order of irrigation pollution > fertilizer pollution. The indicators that were significantly correlated with TP were farm irrigation, fertilizer application, and crop yield, and they all showed a strongly positive correlation. The range of the correlation coefficient was 0.592–0.684, and the correlation degree from small to large was as follows: fertilizer application (0.592) < crop yield (0.660) < farm irrigation (0.684). Hence, the contribution of each pollution source to TP emission was in the order of irrigation pollution > fertilizer pollution. COD had a significantly positive correlation with farm irrigation and crop yield. The range of the correlation coefficient was 0.546–0.576, and the correlation degree from small to large was crop yield (0.546) < farm irrigation (0.576). Thus, irrigation pollution had the greatest influence on COD emissions. Given the continuous improvement of the level of agricultural development, organic or inorganic pollutants, such as nitrogen, phosphorus, and pesticides, enter the surface water, groundwater, and soil environment through surface runoff. Therefore, the agricultural pollution load generated by the basin is also relatively high. All this pollution endangers the environment and the health of the inhabitants [64].

As shown in Figure 10b, at the $p < 0.05$ level, NH_3-N was positively correlated with indicators for animal husbandry and fishery indicators. The range of the correlation coefficient was 0.734–0.940, and the order of the correlation degree from small to large was number of pigs (0.734) < fish breeding (0.811) < number of sheep (0.844) < eggs of poultry (0.849) < number of rabbits (0.916) < number of poultry (0.940). Thus, the contribution of each pollution source to NH_3-N emissions was ranked as pollution from livestock and poultry stocks > pollution from fish breeding. Except for fish breeding, which was not significantly correlated with TP, all other indicators were significantly positively correlated with TP. The correlation coefficients ranged from 0.632 to 0.799, and the order of the correlation degree was number of pigs (0.632) < eggs of poultry (0.664) < number of poultry (0.689) < number of sheep (0.756) < number of rabbits (0.799). The number of sheep, poultry, and rabbits were significantly positively correlated with COD, and the correlation coefficient ranged from 0.677 to 0.769 and were ranked as number of sheep (0.677) < number of poultry (0.701) < number of rabbit (0.769).

As shown in Figure 10c, at the level of $p < 0.05$, NH_3-N had a significantly positive correlation with metallurgical building materials, food and beverage, and electric energy in the industrial indicators. The correlation coefficient ranged from 0.536 to 0.680. The order of the correlation degree from small to large was electric energy (0.536) < food and beverage (0.643) < metallurgical building materials (0.680). Thus, the contribution of each pollution source to NH_3-N emissions was ranked as metallurgical building materials pollution > food and beverage pollution > electric energy pollution. TP was significantly negatively correlated with electricity, heat, gas, and water production and supply. The correlation coefficient was 0.592. Hence, the metallurgical building materials industry had the greatest impact on TP emissions.

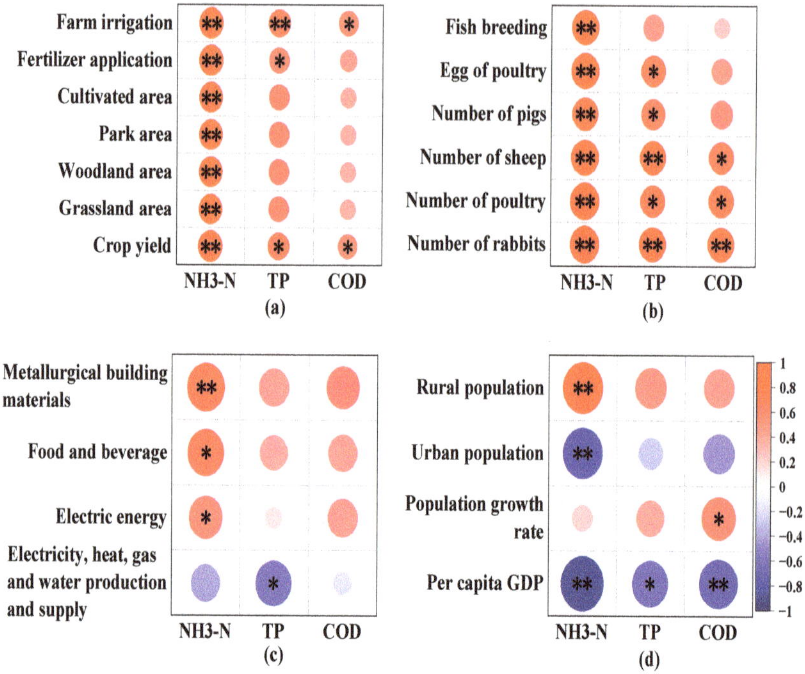

Figure 10. The correlations between pollutants and industry indicators. Industry indicators include agriculture and forestry (**a**), animal husbandry and fishery (**b**), industry (**c**), and population economy (**d**). (* $p \leq 0.05$, ** $p \leq 0.01$).

As shown in Figure 10d, at the $p < 0.05$ level, NH$_3$-N had a highly significantly positive correlation with rural population (0.77), and a highly significantly negative correlation with urban population (−0.75) and per capita GDP (−0.88). TP was significantly negatively correlated with GDP per capita (−0.62). COD was significantly positively correlated with population growth rate (0.55) and extremely significantly negatively correlated with per capita GDP (−0.70). At present, domestic sewage has become one of the most important sources of water pollution in China. The population density of the Tuojiang River Basin is considerably higher than that of other basins; thus, the sewage load generated by urban residents in their lifetime is relatively large [4]. However, the construction of township sewage treatment facilities is not complete, and the collection rate of the pipe network is low. Therefore, sewage treatment facilities and supporting pipe networks in townships and rural gathering points should be improved; and domestic garbage collection, transfer, and treatment systems must be established and improved.

According to the correlation analysis between pollutants and industry indicators, in the order of correlation from strong to weak, NH$_3$-N in the river was correlated with the number of poultry, number of rabbits, crop yield, farm irrigation, fertilizer application, eggs of poultry, number of sheep, cultivated area, park area, woodland area, grassland area, fish breeding, number of pigs, metallurgical building materials, food and beverages, and electric energy. The TP in the river was correlated with the number of rabbits; number of sheep; number of poultry; farm irrigation; eggs of poultry; crop yield; number of pigs; electricity, heat, gas, and water production and supply; and fertilizer application. The COD in the river was correlated with the number of rabbits, number of poultry, number of sheep, farm irrigation, and crop yield. In conclusion, the concentrations of NH_3-N, TP, and COD in the river were mainly related to stock breeding and farmland irrigation.

4. Conclusions

This paper investigated the water environment quality evaluation and pollution source analysis in the Tuojiang River (NJC section) from 2015 to 2019. The results showed a general decline in grey water footprint and also the improvement of water environment quality. Significant variations were determined within the proportion of pollutants in GWF_{Total}, and GWF_{TP} accounted for the largest proportion; thus, the grey water footprint was GWF_{TP}. The main sources of GWF_{COD}, GWF_{NH3-N}, and GWF_{TP} were agricultural pollution. The proportion of agricultural pollution in GWF_{COD} and GWF_{NH3-N} decreased mainly due to the reduction in livestock breeding. The proportion of agricultural pollution in GWF_{TP} increased, which was mainly due to the increase in the area of farm irrigation. The economic development level and water environment of the Tuojiang River (NJC section) were generally in a high-quality coordination state, indicating that the economic development level was relatively high and the damage to the water environment was relatively low. The pollutants exceeding the standard in typical sections were concentrated in the dry season, and the highest rate of TP exceeded the standard. The correlation analysis of the main pollutants showed that the source of pollutants within the Tuojiang River were imported pollution, stock breeding pollution, and farmland irrigation pollution.

In response to the matter of water environment quality and pollution source, the main target is controlling agricultural pollution and upstream imported pollution. To control agricultural source pollution, attention should be given to the problem of direct discharge of wastewater from farms that do not meet the standards. Abuse of chemical fertilizers and pesticides in agricultural production should be avoided, and efficient fertilization and irrigation technology must be promoted. Sewage treatment facilities at township gathering points should also be improved. In view of the problem of upstream input pollution, the pollution discharge problem of upstream phosphogypsum factories should be given attention.

Author Contributions: Conceptualization, K.Z. and S.W.; methodology, K.Z., S.W. and S.L.; software, S.W.; validation, K.Z. and S.W.; formal analysis, S.W. and S.L.; investigation, K.L., J.Y. and X.L.; resources, K.Z. and S.W.; data curation, K.Z. and S.W.; writing—original draft preparation, K.Z. and S.W.; writing—review and editing, K.Z. and S.W.; visualization, K.L.; supervision, J.Y.; project administration, X.L.; funding acquisition, K.Z. All authors have read and agreed to the published version of the manuscript.

Funding: This research was funded by The Xinjiang Talent Introduction Program (2020), the National Natural Science Foundation of China, grant number 42177037, and the Science and Technology and Technology Innovation Projects of Shenhua Shendong Coal Group, grant number 202016000041.

Institutional Review Board Statement: Not applicable.

Informed Consent Statement: Not applicable.

Data Availability Statement: Publicly available datasets were analyzed in this study. This data can be found here: https://www.neijiang.gov.cn/.

Acknowledgments: We acknowledge all the authors for their contributions. We sincerely thank the anonymous reviewers and the editor for their effort to review this manuscript.

Conflicts of Interest: The authors declare no conflict of interest.

References

1. Tian, Y.; Sun, C. Comprehensive carrying capacity, economic growth and the sustainable development of urban areas: A case study of the Yangtze River economic belt. *J. Clean. Prod.* **2018**, *195*, 486–496. [CrossRef]
2. Que, S.S.; Luo, H.Y.; Wang, L.; Zhou, W.Q.; Yuan, S.C. Canonical correlation study on the relationship between shipping development and water environment of the Yangtze River. *Sustainability* **2020**, *12*, 3279. [CrossRef]

3. Chen, Y.S.; Zhang, S.H.; Huang, D.S.; Li, B.L.; Liu, J.G.; Liu, W.J.; Ma, J.; Wang, F.; Wang, Y.; Wu, S.J.; et al. The development of China's Yangtze River Economic Belt: How to make it in a green way? *Sci. Bull.* **2017**, *62*, 648–651. [CrossRef]
4. Liu, D.D.; Bai, L.; Qiao, Q.; Zhang, Y. Anthropogenic total phosphorus emissions to the Tuojiang River Basin, China. *J. Clean. Prod.* **2021**, *294*, 126325. [CrossRef]
5. Zhang, W.Q.; Jin, X.; Cao, H.M.; Zhao, Y.; Shan, B.Q. Water quality in representative Tuojiang river network in southwest China. *Water* **2018**, *10*, 864. [CrossRef]
6. Zhou, G.Y.; Wang, Q.G.; Zhang, J.; Li, Q.S.; Wang, Y.Q.; Wang, M.J.; Huang, X. Distribution and characteristics of microplastics in urban waters of seven cities in the Tuojiang River basin, China. *Environ. Res.* **2020**, *189*, 109893. [CrossRef]
7. Kumar, V.; Parihar, R.D.; Sharma, A.; Bakshi, P.; Sidhu, G.P.S.; Bali, A.S.; Karaouzas, L.; Bhardwaj, R.; Thukral, A.K.; Gyasi-Agyei, Y.; et al. Global evaluation of heavy metal content in surface water bodies: A meta-analysis using heavy metal pollution indices and multivariate statistical analyses. *Chemosphere* **2019**, *236*, 124364. [CrossRef] [PubMed]
8. Fan, C.Z.; Liu, Y.B.; Liu, C.H.; Zhao, W.B.; Hao, N.X.; Guo, W.; Yuan, J.H.; Zhao, J.J. Water quality characteristics, sources, and assessment of surface water in an industrial mining city, southwest of China. *Environ. Monit. Assess.* **2022**, *194*, 259. [CrossRef]
9. Maier, H.R.; Dandy, G.C. Neural networks for the prediction and forecasting of water resources variables a review of modelling issues and applications. *Environ. Modell. Softw.* **2000**, *15*, 101–124. [CrossRef]
10. Jang, D.; Choi, G. Estimation of Non-Revenue Water Ratio for Sustainable Management Using Artificial Neural Network and Z-Score in Incheon, Republic of Korea. *Sustainability* **2017**, *9*, 1933. [CrossRef]
11. Tharme, R.E. A global perspective on environmental flow assessment: Emerging trends in the development and application of environmental flow methodologies for rivers. *River Res. Appl.* **2003**, *19*, 397–441. [CrossRef]
12. Xia, X.H.; Zhang, X.; Yang, Z.F.; Shen, Z.Y.; Dong, L.I. Integrated water quality and quantity evaluation of the Yellow River. *J. Nat. Resour.* **2004**, *19*, 293–299. [CrossRef]
13. Wang, X.Q.; Liu, C.M.; Zhang, Y. Water quantity/quality combined evaluation method for rivers' water requirements of the instream environmental flow in dualistic water cycle: A case study of Liaohe river basin. *J. Geogr. Sci.* **2007**, *3*, 304–316. [CrossRef]
14. Hoekstra, A.Y.; Mekonnen, M.M. The water footprint of humanity. *Proc. Nat. Acad. Sci. USA* **2012**, *109*, 3232–3237.
15. Ansorge, L.; Stejskalová, L.; Dlabal, J. Grey water footprint as a tool for implementing the Water Framework Directive-Temelín nuclear power station. *J. Clean. Prod.* **2020**, *263*, 121541. [CrossRef]
16. Miglietta, P.P.; Toma, P.; Fanizzi, F.P.; De Donno, A.; Coluccia, B.; Migoni, D.; Bagordo, F.; Serio, F. A Grey Water Footprint Assessment of Groundwater Chemical Pollution: Case Study in Salento (Southern Italy). *Sustainability* **2017**, *9*, 799.
17. Allocca, V.; Marzano, E.; Tramontano, M.; Celico, F. Environmental impact of cattle grazing on a karst aquifer in the southern Apennines (Italy): Quantification through the grey water footprint. *Ecol. Indic.* **2018**, *93*, 830–837.
18. Hu, Y.C.; Huang, Y.F.; Tang, J.X.; Gao, B.; Yang, M.H. Evaluating agricultural grey water footprint with modeled nitrogen emission data. *Resour. Conserv. Recycl.* **2018**, *138*, 64–73.
19. Muratoglu, A. Grey water footprint of agricultural production: An assessment based on nitrogen surplus and high-resolution leaching runoff fractions in Turkey. *Sci. Total Environ.* **2020**, *742*, 140553. [PubMed]
20. Zhao, D.D.; Tang, Y.; Liu, J.G.; Tillotson, M.R. Water footprint of Jing-Jin-Ji urban agglomeration in China. *J. Clean. Prod.* **2017**, *167*, 919–928.
21. Johnson, M.B.; Mehrvar, M. An assessment of the grey water footprint of winery wastewater in the Niagara region of Ontario, Canada. *J. Clean. Prod.* **2019**, *214*, 623–632. [CrossRef]
22. Ma, X.T.; Yang, D.L.; Shen, X.X.; Zhai, Y.J.; Zhang, R.R.; Hong, J.L. How much water is required for coal power generation: An analysis of gray and blue water footprints. *Sci. Total Environ.* **2018**, *636*, 547–557. [CrossRef] [PubMed]
23. Martinez-Alcala, I.; Pellicer-Martinez, F.; Fernandez-Lopez, C. Pharmaceutical grey water footprint: Accounting, influence of wastewater treatment plants and implications of the reuse. *Water Res.* **2018**, *135*, 278–287. [CrossRef] [PubMed]
24. Cai, B.; Liu, B.; Zhang, B. Evolution of Chinese urban household's water footprint. *J. Clean. Prod.* **2019**, *208*, 1–10. [CrossRef]
25. Chukalla, A.D.; Krol, M.S.; Hoekstra, A.Y. Trade-off between blue and grey water footprint of crop production at different nitrogen application rates under various field management practices. *Sci. Total Environ.* **2018**, *626*, 962–970. [CrossRef]
26. Mekonnen, M.M.; Hoekstra, A.Y. Global gray water footprint and water pollution levels related to anthropogenic nitrogen loads to fresh water. *Environ. Sci. Technol.* **2015**, *49*, 12860–12868. [CrossRef]
27. Mekonnen, M.M.; Lutter, S.; Martinez, A. Anthropogenic nitrogen and phosphorus emissions and related grey water footprints caused by EU-27's crop production and consumption. *Water* **2016**, *8*, 30. [CrossRef]
28. Shrestha, S.; Pandey, V.P.; Chanamai, C.; Ghosh, D.K. Green, blue and grey water footprints of primary crops production in Nepal. *Water Resour. Manag.* **2014**, *27*, 5223–5243. [CrossRef]
29. Wan, L.Y.; Cai, W.J.; Jiang, Y.K.; Wang, C. Impacts on quality-induced water scarcity: Drivers of nitrogen-related water pollution transfer under globalization from 1995 to 2009. *Environ. Res. Lett.* **2016**, *11*, 074017. [CrossRef]
30. Li, H.; Liu, G.Y.; Yang, Z.F. Improved gray water footprint calculation method based on a mass-balance model and on fuzzy synthetic evaluation. *J. Clean. Prod.* **2019**, *219*, 377–390. [CrossRef]
31. Feng, H.; Sun, F.; Liu, Y.; Zeng, P.; Che, Y. Mapping multiple water pollutants across China using the grey water footprint. *Sci. Total Environ.* **2021**, *785*, 147255. [CrossRef] [PubMed]

32. Jiao, S.X.; Wang, A.Z.; Chen, L.F.; Chen, J.W.; Zhang, X.X.; Zhao, R.Q.; Zhang, C.C. Study on the decoupling relationship between the economic development and water environment in Henan province based on the grey water footprint theory. *Innov. Sci. Technol.* **2018**, *18*, 32–36.
33. Lertpaitoonpan, W.; Ong, S.; Moorman, T. Effect of organic carbon and pH on soil sorption of sulfamethazine. *Chemosphere* **2009**, *76*, 558–564. [CrossRef]
34. Zhang, K.; Gao, J.; Jiang, B.; Han, J.; Chen, M. Experimental study on the mechanism of water-rock interaction in the coal mine underground reservoir. *Meitan Xuebao* **2019**, *44*, 3760–3772.
35. Guo, Y.Z.; Wang, X.Y.; Zhou, L.; Melching, C.; Li, Z.Q. Identification of Critical Source Areas of Nitrogen Load in the Miyun Reservoir Watershed under Different Hydrological Conditions. *Sustainability* **2020**, *12*, 964. [CrossRef]
36. Zhang, K.; Li, X.N.; Song, Z.Y.; Yan, J.Y.; Chen, M.Y.; Yin, J.C. Human health risk distribution and safety threshold of Cadmium in soil of coal chemical industry area. *Minerals* **2021**, *11*, 678. [CrossRef]
37. Li, J.; Zhang, H.; Chen, Y.; Luo, Y.; Zhang, H. Sources identification of antibiotic pollution combining land use information and multivariate statistics. *Environ. Monit. Assess.* **2016**, *188*, 430. [CrossRef] [PubMed]
38. Ji, X.N.; Chen, J.H.; Guo, Y.L. A Multi-Dimensional Investigation on Water Quality of Urban Rivers with Emphasis on Implications for the Optimization of Monitoring Strategy. *Sustainability* **2022**, *14*, 4174. [CrossRef]
39. Li, L.; Liu, D.; Zhang, Q.; Song, K.; Zhou, X. Occurrence and ecological risk assessment of selected antibiotics in the freshwater lakes along the middle and lower reaches of Yangtze River Basin. *J. Environ. Manag.* **2019**, *249*, 109396. [CrossRef]
40. Zhang, K.; Zheng, X.H.; Li, X.N.; Li, J.H.; Ji, Y.N.; Gao, H. The pollution assessment and source analysis of groundwater in a region of southwest China. *J. Henan Norm. Univ. (Nat. Sci. Ed.)* **2020**, *48*, 64–73.
41. Kang, W.; Lin, Z.; Zhang, R. Impact of phosphate mining and separation of mined materials on the hydrology and water environment of the Huangbai River basin, China. *Sci. Total Environ.* **2016**, *543*, 347–356.
42. Zhang, M.; Chen, X.L.; Yang, S.H.; Song, Z.; Wang, Y.G.; Yu, Q. Basin-scale pollution loads analyzed based on coupled empirical models and numerical models. *Int. J. Environ. Res. Public Health* **2021**, *18*, 12481. [CrossRef] [PubMed]
43. Zeng, Z.; Luo, W.G.; Wang, Z.; Yi, F.C. Water pollution and its causes in the Tuojiang river basin, China: An artificial neural network analysis. *Sustainability* **2021**, *13*, 792. [CrossRef]
44. Wu, B.; Zeng, W.H.; Chen, H.H.; Zhao, Y. Grey water footprint combined with ecological network analysis for assessing regional water quality metabolism. *J. Clean. Prod.* **2016**, *112*, 3138–3151. [CrossRef]
45. Ang, B.W. Decomposition analysis for policymaking in energy: Which is the preferred method? *Energy Policy* **2004**, *32*, 1131–1139. [CrossRef]
46. Li, J.; Hou, L.P.; Wang, L.; Tang, L.N. Decoupling Analysis between Economic Growth and Air Pollution in Key Regions of Air Pollution Control in China. *Sustainability* **2021**, *13*, 6600. [CrossRef]
47. Cahill, S.A. Calculating the rate of decoupling for crops under cap/oilseeds reform. *J. Agric. Econ.* **1997**, *48*, 349–378. [CrossRef]
48. Tapio, P. Towards a theory of decoupling: Degrees of decoupling in the EU and the case of road traffic in Finland between 1970 and 2001. *Transp. Policy* **2005**, *12*, 137–151. [CrossRef]
49. Vehmas, J.; Luukkanen, J.; Kaivo-Oja, J. Linking analyses and environmental Kuznets curves for aggregated material flows in the EU. *J. Clean. Prod.* **2007**, *15*, 1662–1673. [CrossRef]
50. Her, Q.L.; Wong, J. Significant correlation versus strength of correlation. *Am. J. Health-Syst. Pharm.* **2020**, *77*, 73–75. [CrossRef]
51. Schober, P.; Boer, C.; Schwarte, L.A. Correlation coefficients: Appropriate use and interpretation. *Anesth. Analg.* **2018**, *126*, 1763–1768. [CrossRef] [PubMed]
52. Edelmann, D.; Mori, T.F.; Szekely, G.J. On relationships between the Pearson and the distance correlation coefficients. *Stat. Probab. Lett.* **2021**, *169*, 108960. [CrossRef]
53. Cui, S.B.; Dong, H.J.; Wilson, J. Grey water footprint evaluation and driving force analysis of eight economic regions in China. *Environ. Sci. Pollut. Res.* **2020**, *27*, 20380–20391. [CrossRef] [PubMed]
54. Jamshidi, S.; Imani, S.; Delavar, M. An approach to quantifying the grey water footprint of agricultural productions in basins with impaired environment. *J. Hydrol.* **2022**, *606*, 127458. [CrossRef]
55. Hu, Y.Y.; Wang, Y.D.; Li, T.X.; Zheng, Z.C.; Pu, Y. Characteristics analysis of agricultural non-point source pollution on Tuojiang River basin. *Sci. Agric. Sin.* **2015**, *48*, 3654–3665.
56. Yao, J.; Yang, L.J.; Xiao, Y.T.; Fan, M.; Zhan, S.; Liu, Y.F.; Wang, H.W.; Chen, W.; Deng, Y.; Wang, M.L. Spatial-temporal evolution of agricultural non-point sources of total phosphorus pollution loads in Tuojiang River watershed based on correction of social-economic factors. *J. Agro-Environ. Sci.* **2022**, *41*, 1022–1035.
57. Meng, W.; Feagin, R.A. Mariculture is a double-edged sword in China. *Estuar. Coast. Shelf Sci.* **2019**, *222*, 147–150. [CrossRef]
58. Cai, C.F.; Gu, X.H.; Ye, Y.T.; Yang, C.G.; Dai, X.Y.; Chen, D.X.; Yang, C. Assessment of pollutant loads discharged from aquaculture ponds around Taihu Lake, China. *Aquac. Res.* **2011**, *44*, 795–806. [CrossRef]
59. Wang, H.; Xu, Y.L.; Zhang, Q.; Lin, C.W.; Zhai, L.M.; Liu, H.T.; Pu, B. Emission characteristics of nitrogen and phosphorus in a typical agricultural small watershed in Tuojiang river basin. *Environ. Sci.* **2020**, *41*, 4547–4554.
60. Cao, Y. Analysis of water quality of backwater of farmland irrigation and its influence on examination section. *Environ. Sci. Technol.* **2021**, *34*, 67–71.
61. Zhang, K.; Bai, L.; Wang, P.F.; Zhu, Z. Field measurement and numerical modelling study on mining-induced subsidence in a typical underground mining area of northwestern China. *Adv. Civ. Eng.* **2021**, *2021*, 5599925. [CrossRef]

62. Zhang, K.; Gao, J.; Men, D.; Zhao, X.; Wu, S. Insight into the heavy metal binding properties of dissolved organic matter in mine water affected by water-rock interaction of coal seam goaf. *Chemosphere* **2021**, *265*, 129134. [CrossRef] [PubMed]
63. Ma, Y.Q.; Qin, Y.W.; Zheng, B.H.; Zhang, L.; Zhao, Y.M. Seasonal variation of enrichment, accumulation and sources of heavy metals in suspended particulate matter and surface sediments in the Daliao river and Daliao river estuary, Northeast China. *Environ. Earth Sci.* **2015**, *73*, 5107–5117. [CrossRef]
64. Li, S.P.; Gong, Q.X.; Yang, S.L. Analysis of the agricultural economy and agricultural pollution using the decoupling index in Chengdu, China. *Int. J. Environ. Res. Public Health* **2019**, *16*, 4233. [CrossRef] [PubMed]

Article

Water Chemical Characteristics and Safety Assessment of Irrigation Water in the Northern Part of Hulunbeier City, Grassland Area in Eastern China

Wanli Su [1,2], Feisheng Feng [1,3,*], Ke Yang [1,3,*], Yong Zhou [1,3], Jiqiang Zhang [1,3] and Jie Sun [1,3]

1 State Key Laboratory of Mining Response and Disaster Prevention and Control in Deep Coal Mines, Anhui University of Science & Technology, Huainan 232001, China
2 Institute of Coal Chemical Industry Technology, China Energy Group, Ningxia Coal Industry Co., Ltd., Yinchuan 750411, China
3 Institute of Energy, Hefei Comprehensive National Science Center, Hefei 230031, China
* Correspondence: fengfeisheng21@aust.edu.cn (F.F.); yksp2003@163.com (K.Y.)

Citation: Su, W.; Feng, F.; Yang, K.; Zhou, Y.; Zhang, J.; Sun, J. Water Chemical Characteristics and Safety Assessment of Irrigation Water in the Northern Part of Hulunbeier City, Grassland Area in Eastern China. *Sustainability* 2022, 14, 16068. https://doi.org/10.3390/su142316068

Academic Editor: Ning Yuan

Received: 27 September 2022
Accepted: 25 November 2022
Published: 1 December 2022

Publisher's Note: MDPI stays neutral with regard to jurisdictional claims in published maps and institutional affiliations.

Copyright: © 2022 by the authors. Licensee MDPI, Basel, Switzerland. This article is an open access article distributed under the terms and conditions of the Creative Commons Attribution (CC BY) license (https://creativecommons.org/licenses/by/4.0/).

Abstract: Hulun Buir Grassland is a world-famous natural pasture. The Chenbalhu Banner coalfield, the hinterland of the grassland, is located on the west slope of the Great Khingan Mountains and on the north bank of the Hailar River in China. The proven geological reserves of coal are 17 billion tons. Hulun Buir Grassland plays a role in the ecological barrier, regional coal industry, power transmission from west to east and power transmission from north to south. The proportion of local groundwater in irrigation, domestic and industrial production water sources is about 86%. The large-scale exploitation of coal resources and the continuous emergence of large unit and coal-fired power plants have consumed a large amount of local water resources, resulting in the decrease of the local groundwater level and changing the natural flow field of groundwater. This paper studies the background hydrochemical values and evaluates the irrigatibility of the whole Chenbaerhu Banner coalfield, and studies the impact of coal industry chains such as mining areas and coal chemical plants on the hydrochemistry characteristics of groundwater. The above two studies provide important guiding values for guiding local economic structure planning, groundwater resources exploitation and ecological governance. The study found that Na^+ and HCO_3^- in the groundwater in the study area occupy a dominant position. Referring to the comparison of the lowest values of three types of water standards in the Quality Standards for Groundwater (GB/T14848-2017), the amount of NH_4^+, Na^+ and NO_2^- exceeding the standard is close to more than 30%. The main chemical types of river water in the study area are HCO_3^- Na and HCO_3^- Ca·Na, the main chemical types of surface water are HCO_3^- Na and HCO_3^- Na·Ca, and the main chemical type of confined water is HCO_3^- Na. The formation of hydrochemical types is mainly affected by the dissolution, filtration and evaporation of rocks, specifically the dissolution and filtration of sodium and calcium salts. The chemical correlation analysis of groundwater shows that there are abnormal values at many points in the study area. Further combining with the horizontal comparison of surface human activities in the study area, it shows that the influence scope of coal mine production and coal chemical plants on groundwater is extremely limited. The local groundwater is mainly polluted by a large quantity of local cattle and sheep manure, industrial and domestic sewage pollution and farmland fertilizer.

Keywords: groundwater; high-intensity mining; irrigation grade; chemical properties; ecological management

1. Introduction

Groundwater is an important part of water resources. Due to its stable and good water chemistry characteristics, it is one of the important water sources for agricultural irrigation, industrial mining and urban development. In recent years, a growing number of scholars have recognized that groundwater research that monitors groundwater chemistry characteristics and quantity, classifies hydrochemical characteristics, meets the growing demand and evaluates groundwater chemistry characteristics has become crucial. Many scholars at home and abroad have conducted in-depth studies on groundwater chemistry characteristics, hydrochemical classification and water chemistry characteristics evaluation in typical areas by means of testing and analysis, mathematical methods, digital modeling analysis and statistical analysis. Turkish scholar [1] et al. conducted physical and chemical analysis on the water samples collected through the flow path of the Aksu River and used the water quality Index (WQI) method to evaluate the water chemistry characteristics and applicability of drinking water. Indian scholar Narsimha Adimalla [2] studied the quality of water samples collected from the rock-dominated semi-arid region of central Telangana, and analyzed pH, electrical conductivity (EC), total dissolved solids (TDS) and total hardness (TH), calcium (Ca^{2+}), magnesium (Mg^{2+}), sodium (Na^+), potassium (K^+), chloride (Cl^-), sulfate (SO_4^{2-}), nitrate (NO^{3-}) and fluoride (F^-) to evaluate the groundwater drinking category and groundwater irrigation feasibility of excellence. Tanzanian scholar Juma R. Seleman [3] et al. evaluated spatial and seasonal hydrochemical changes, chemical weathering, and hydrological cycles by tracing stable isotopes of $\delta^{18}O$, δ^2H and $^{87}Sr/^{86}Sr$ and dissolved major ions in the Pangani Basin (PRB). Varol Simge [4] and Aniekan Edet [5] applied the water quality index (WQI) to the drinking water assessment system, and respectively determined the applicability of drinking and irrigation uses in the study area, and the source of ions in the river, and obtained relatively scientific evaluation results. K. Arumugam [6] et al. collected contaminated groundwater samples from 62 sites in Tirupur district, Coimbatore district, Tamil Nadu, India, conducted hydrochemical characteristics and groundwater chemistry characteristics assessment, analyzed the main cations and anions, found, using the Piper trilinear chart, that most areas were contaminated with higher concentrations of EC, TDS, K and NO_3, and assessed groundwater for drinking and land irrigation according to the US salinity chart. I. Chenini [7] et al. combined multiple linear regression and structural equation models to analyze the hydrochemical data, and provided a method of characterizing the groundwater chemical characteristics by statistical analysis and modeling of the hydrochemical data of The Mackenzie Basin, so as to explain the chemical origin of groundwater. Lisa M. Galbraith [8] et al. measured the physical variables and nutritional chemistry of 45 water bodies representing a large range of lenticular wetland environments in Otago, New Zealand, and associated them with catchment variables and land, derived watershed boundaries and land coverage from maps, and found that the concentrations of nutrients and other water chemistry characteristics components were positively correlated with the nature and strength of catchment modification. They undertook this study to assess the potential impact of catchment modification on water chemistry characteristics in these different wetlands. Y. Srinivasa Rao [9] et al. collected water samples from existing wells in the Niva River Basin in the Chittoor district of Andhra Pradesh, India, analyzed the main ions and performed a multiple regression analysis to assess the quality of groundwater related to agricultural and household use in the Archea granite and gneiss consortium of Peninsula India. S. Fauriel and L. Laloui [10] developed a general mathematical model to describe the injection, the distribution and the reaction processes of biogrout within a saturated, deformable porous medium. The model was able to reproduce all of the mechanisms of interest and their couplings. Jinhyun Choo and WaiChing Sun [11] developed a theoretical and computational framework for modeling the crystallization-induced deformation and fracture in fluid-infiltrated porous materials. Conservation laws were formulated for coupled chemo–hydro–mechanical processes in a multiphase material composed of the solid matrix, liquid solution, gas, and crystals. The chemo–hydro–mechanical model was coupled with a phase-field approach to fracture which enables simulation of complex fractures without explicitly tracking their geometry. They demonstrated the capability of the proposed modeling framework for simulating complex interactions among unsaturated flow, crystallization kinetics, and cracking in the solid matrix. Javad Ghorbani et al. [12] introduced and numerically implemented a thermos–elasto–plastic constitutive model. They presented and discussed the results of numerical simulations of the performance of GCLs, and identified the effects of deviations from elastic behavior. They conducted a parametric study, which was to identify the influence of key parameters in the constitutive model on the horizontal stress in the GCL's bentonite during restrained shrinkage driven by thermal dehydration.

In response to the problem of over-exploitation of groundwater resources in arid and semi-arid regions of China, many scholars have focused on the impact of pollutants on groundwater and the comprehensive evaluation of groundwater for human health and land irrigation. Peng Suping [13] et al. took the Shengli Coal Mine in Beidang, Xilinhot, as an example, studied and analyzed the hydrochemical characteristics around large open-pit coal mines in the grassland area, and evaluated and analyzed the level of groundwater irrigation in this area. Tang Kewang [14] et al. discussed the groundwater hydrochemical characteristics from four aspects, including spatial variation, salinity distribution, hardness distribution and acid–alkali distribution, and made a comprehensive evaluation of the national water resources with respect to the groundwater chemistry characteristics status. Xiaodong He et al. [15] collected and analyzed groundwater samples from the water-bearing stratum of Luohe in Wuqi County, northwest China. They analyzed the hydrochemical phase of groundwater by statistical analysis and trilinear diagram, studied the natural evolution mechanism of groundwater and surface water via the Gibbs diagram, correlation analysis and the binary diagram, and evaluated the harm to human health. Through hydrogeological investigation and water sample collection, Cao Guangyuan et al. [16] analyzed the spatial distribution characteristics and causes of groundwater water chemistry in Chengjiaying Basin through the groundwater flow system, Gibbs map and Piper. Min Xiao [17] investigated the spatio–temporal variation of hydrochemistry in a typical karst groundwater system in southwest China during the rainy season. Through studying the main ions and $\delta^{13}CDIC$ to track the evolution of carbonate in freshwater aquifer, the biogeochemical processes and the temporal characteristics, they found that coupling analysis of $\delta^{13}CDIC$ and hydrochemical parameters is an effective way to explore the biogeochemical processes of carbon and to track the source of groundwater contaminants in karst regions.

In conclusion, it can be found that most of the hydrochemical types and water chemistry characteristics assessments are focused on the river, the basin, wetland, etc., and relatively few studies have been conducted on the impact of coal industrial chains such as mining areas and coal chemical plants on groundwater hydrochemical characteristics. Farhad Howladar [18] et al. conducted water chemistry characteristics assessments around livestock, drinking water, irrigation purposes and environmental impacts in the Barapukuria coal mine industrial zone in Dinajpur, Bangladesh, and used field investigation, laboratory chemical analysis, statistical representation and the correlation matrix to prove the applicability of laboratory analysis. Jae Gon Kim [19] et al. studied the chemical characteristics of stream water and sediments in a small watershed with two unique mineralized regions (Cu and Pb-Zn), seven abandoned mines and an active quarry to study the influence of mining activities and regional geology on chemistry. Dong-lin [20] et al. collected 76 typical water samples in Yulin City for particle quality tests and a water chemistry characteristics investigation. Using the Romani classification method and principal component analysis, he believed that the groundwater environment in this area largely depends on the characteristic components of the natural groundwater background. Some of the water was polluted by the solid waste produced by leaching coal charcoal mining, and the other part was polluted by the acid mine water from coal seams and improper irrigation, geological and hydrogeological conditions also cause changes in the water environment. Biao Zhang et al. [21] investigated the hydrogeochemical characteristics and groundwater evolution in the Delingha area in the northeast of the Qaidam Basin in northwest China. Through the analysis of collected water samples, they believed that the chemical evolution of groundwater was mainly controlled by the dissolution of evaporites and carbonate minerals, the weathering of aluminosilicates and cation exchange.

The above research has provided guidance for the follow-up study of groundwater chemical characteristics in terms of research methods, research ideas and data processing methods. However, the Chenbaerhu Banner coalfield is located on the west slope of the Greater Hinggan Mountains and the north bank of the Hailar River, spanning the forest grassland and the arid grassland. It has the only pure natural meadow grassland in the world, and its coal geological reserves have been proved to be 17 billion tons. The coupling relationship between the impacts of large grasslands, large coal bases, pastures and coal chemical plants on groundwater is complex, and groundwater is the main water source for local irrigation, living and industrial production. Therefore, analyzing the characteristics and evolution of groundwater in Chenbaerhu Banner coalfield, conducting hydrochemical characteristics research and irrigation grade evaluation are all of great significance to the rational management of groundwater resources in the area, normal and safe mining, ecological recovery after mining, environmental governance, etc.

2. Materials and Methods

Survey Region

The study area is located in the north of the Hulun Buir urban area, including the northeast of Chen Balhu Banner and the west of Hailar, and is located in the hinterland of Hulun Buir Prairie. The Morigel River and Hailar River in the west form a hydrogeological unit. The southern and northwest boundaries are at the junction of the middle and low hills and the Greater Hinggan Mountains, which is a semi-arid continental climate. The annual average precipitation is 350–400 mm, 70% is concentrated in the hot and rainy June–August. The annual average evaporation is 1371.1 mm, and 54% is concentrated in May–July.

Since there is no systematic long-term dynamic observation network of groundwater environment in the study area, we can use drill holes and water samples to conduct field research on the hydrogeological conditions in the study area. Finally, we selected 41 measuring points, 3 water samples for each measuring point, 123 water samples in total, including 1 sewage sample, 3 river water samples, 13 confined water samples and 24 phreatic water samples, as shown in Figure 1. Using the German company OTT Hydrolab multi-parameter water chemistry characteristics monitor [22], parameters such as pH, EC, COD, total nitrogen, NH_4^+ and conductivity of water samples were directly measured on site [23–25]. In the laboratory, ICP-MS (Agilent-7700x, Palo Alto, CA, USA) of Agilent Company was used to measure the concentration of trace elements [26], Dionex DX-500 ion chromatography system was used to measure anions, and the Skalar San++continuous flow injection analysis system of the Netherlands was used to quantitatively measure the concentration of total nitrogen and ammonium nitrogen [27–29]. The Hydrolab multi parameter water chemistry characteristics monitor adopts electrical plus photometric measurement, which is characterized by easy operation, convenient storage and transportation, and accurate results. It is suitable for rapid detection of various physical and chemical parameters and ion concentration in water on site. The portable water chemistry characteristics monitor is easy to carry and can be used for real-time continuous detection. It can provide long-term continuous real-time data to judge the current situation and change trend of water chemistry characteristics in the surveyed area. Inductively coupled plasma mass spectrometry (ICP-MS) is used for testing. This method can have a small measurement limit in the detection process of metal ions, high accuracy of detection results, and a small matrix effect, which can meet the requirements of joint detection of multiple elements. Therefore, it has a wider application limit in practical applications. The Skala San++continuous flow injection analysis system uses the continuous flow method to compress elastic pump pipes with different inner diameters through a peristaltic pump. The reagent and sample are drawn into the pipeline system in proportion, and air bubbles are introduced as sample intervals to make the reagent and sample mix and react under certain conditions. After color development, the reagent and sample were put into the colorimetric cell to measure the absorbance. According to Lambert Beer's law, the sample concentration is measured quantitatively. After data transmission and the processing system, the analysis results are obtained.

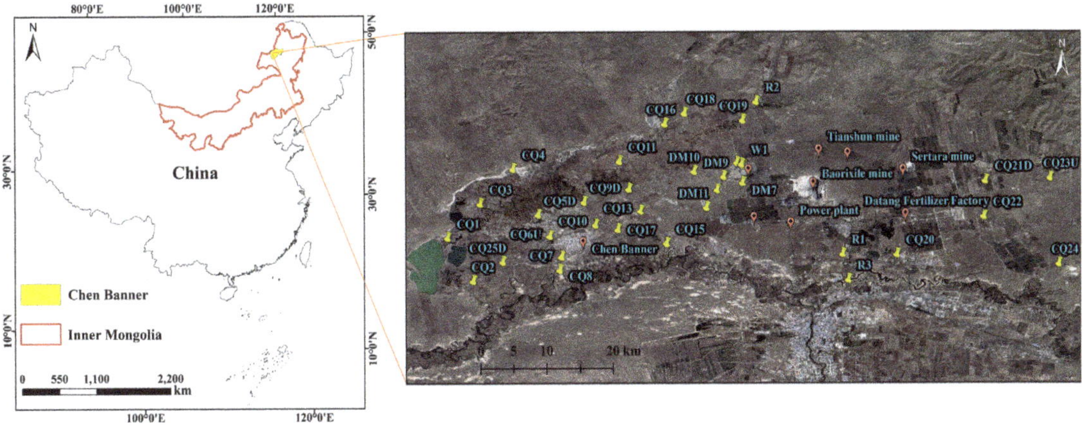

Figure 1. Chenbaerhu Banner coalfield geographical location map; groundwater sampling and ion analysis.

3. Results and Discussion

3.1. Descriptive Statistical Analysis of Groundwater Chemistry

The proportion relationship of ion content in the water samples is shown in Figure 2. The proportion relationship of ion content in the water samples has obvious characteristics. In 41 sample cations, except DM4, DM10, CQ15, CQ10, CQ7, R3 and R2, the concentration of cations in each remaining water sample is $Na^+ > Ca^{2+} > Mg^{2+} > K^+$. In DM4, DM10, CQ15, CQ10, CQ7, R3 and R2, the concentration of cations is $Ca^{2+} > Na^+ > Mg^{2+} > K^+$. In 41 samples, the concentration of cations is $Ca^{2+} > Na^+ > Mg^{2+} > K^+$. The average concentrations of Na^+, Ca^{2+}, Mg^{2+} and K^+ ions in the 41 samples were 157.27 mg/L, 49.05 mg/L, 25.80 mg/L and 3.46 mg/L. In addition, among cations, Na^+ and Ca^{2+} are the main components in the groundwater in the study area, and the content ranges from 6.01 to 501.2 mg/L and 12.02 to 194.39 mg/L. The variation coefficients of Na^+ and Ca^{2+} are large.

Among the anions in the 41 samples, HCO_3^- occupies an absolute dominant position. Except for the two measuring points CQ20 and CQ22, HCO_3^- accounts for more than half of the anions in all the remaining measuring points, and 11 measuring points account for more than 90% of the anions. The average concentrations of HCO_3^-, SO_4^{2-}, Cl^- and NO_3^- ions in the 41 samples are 524.09 mg/L, 76.18 mg/L, 98.30 mg/L and 1.18 mg/L, respectively. Compared with the Class 3 standards in the Quality Standard for Groundwater [30] (GB/T14848-2017), NH_4^+, Na^+, NO_2^-, TDS, Cl^- and SO_4^{2-} were found to exceed the standard excessively. The number of excessive ions is small, and other ions are within a reasonable range. Among them, there are 14 water samples with NH_4^+ exceeding its standard value by 0.2 mg/L, and the exceeding rate of the standard is 34.15%; there are 12 water samples with Na^+ exceeding its standard value by 200 mg/L, and the exceeding rate of the standard is 29.27%; there are 12 water samples with NO_2^- exceeding its standard value by 0.02 mg/L, and the exceeding rate of the standard is 29.27%; there are 4 water samples with TDS exceeding its standard value by 1000 mg/L, and the exceeding rate of the standard is 9.76%; there are 3 water samples with Cl^- exceeding its standard value by 250 mg/L, and the exceeding rate of the standard is 7.32%.

3.2. Chemical Correlation Analysis of Groundwater

The above analysis can find that there are some ions with large coefficient of variation in the water pattern, which indicates that they are sensitive ions that change with environmental factors. However, it is unscientific to determine whether they are the categories with large variability simply through the magnitude of numerical variation. The occurrence of abnormal values and the analysis of their causes are also important opportunities to study the evolution characteristics of groundwater in the region. Therefore, based on the above analysis, the concept of box graph [31] is introduced to find the abnormal values of various ions in water; the box graph can calculate the upper edge, the upper quartile Q3, the median, the lower quartile Q1 and the lower edge, and all apart from the upper and lower edges of a group of data are abnormal values, as shown in Figure 3, which is the box graph of anions and cations.

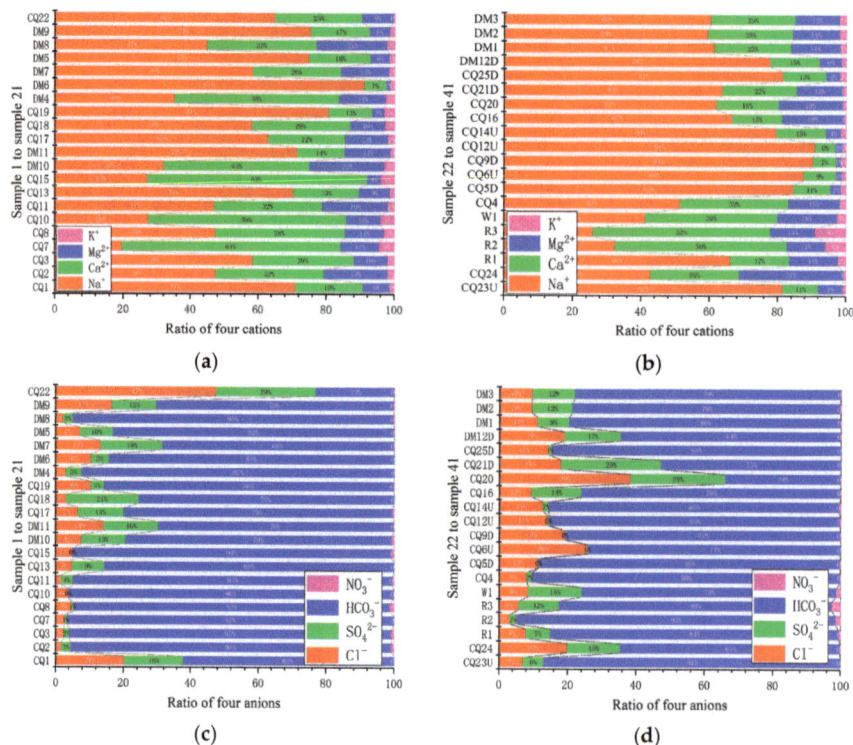

Figure 2. Groundwater chemical specific gravity diagram of main anion and anion in 41 samples. (**a**) Sample 1–21 Specific gravity of the four cations. (**b**) Sample 2–41 Specific gravity of the four cations. (**c**) Sample 1–21 Specific gravity of the four anions. (**d**) Sample 2–41 Specific gravity of the four anions.

By comparing the background values of India [32–34], the United States [35,36], Iran [37,38] and Xilinhot City [39,40] near the study area, it can be found that the concentration of Na^+ at the measuring point CQ22 is abnormally large, with the concentration value of 501.2 mg/L. The concentration of Ca^{2+} at the measuring point CQ22 is abnormally large, with the concentration value of 194.39 mg/L. The concentrations of Mg^{2+} at the measuring points CQ20 and CQ24 are abnormally large, with the concentration values of 107.45 mg/L and 91.89 mg/L, respectively. The concentration of Cl^- values at the measuring points CQ22, CQ20 and CQ23U are abnormally large, with the concentration values of 779.77 mg/L, 514.07 mg/L and 343.89 mg/L, respectively. The concentration of SO_4^{2-} values of measuring points CQ22, CQ20 and CQ23U are abnormally large, with the concentration values of 483.2 mg/L, 368.89 mg/L and 297.8 mg/L, respectively. The concentration of HCO_3^- at the measuring point CQ23U is abnormally large, with the concentration value of 4210.2 mg/L. The concentration of NO_3^- at the measuring points W1,

CQ2 and CQ3 are abnormally large, with the concentration values of 5.8 mg/L, 4.86 mg/L and 4.14 mg/L, respectively. The concentration of NO_2^- at the measuring point DM6, R1, DM5 and DM7 are abnormally large, with the concentration values of 1.68 mg/L, 0.48 mg/L, 0.26 mg/L and 0.12 mg/L, respectively. The concentration of F^- in measuring points DM8, CQ16, R1 and DM11 are abnormally large, with the concentration values of 1.28 mg/L, 1.27 mg/L, 1.23 mg/L and 1.12 mg/L, respectively. The concentration of H_2SiO_3 at measuring points W1 and CQ9D are abnormally large, with the concentration value of 47.58 mg/ m², respectively.

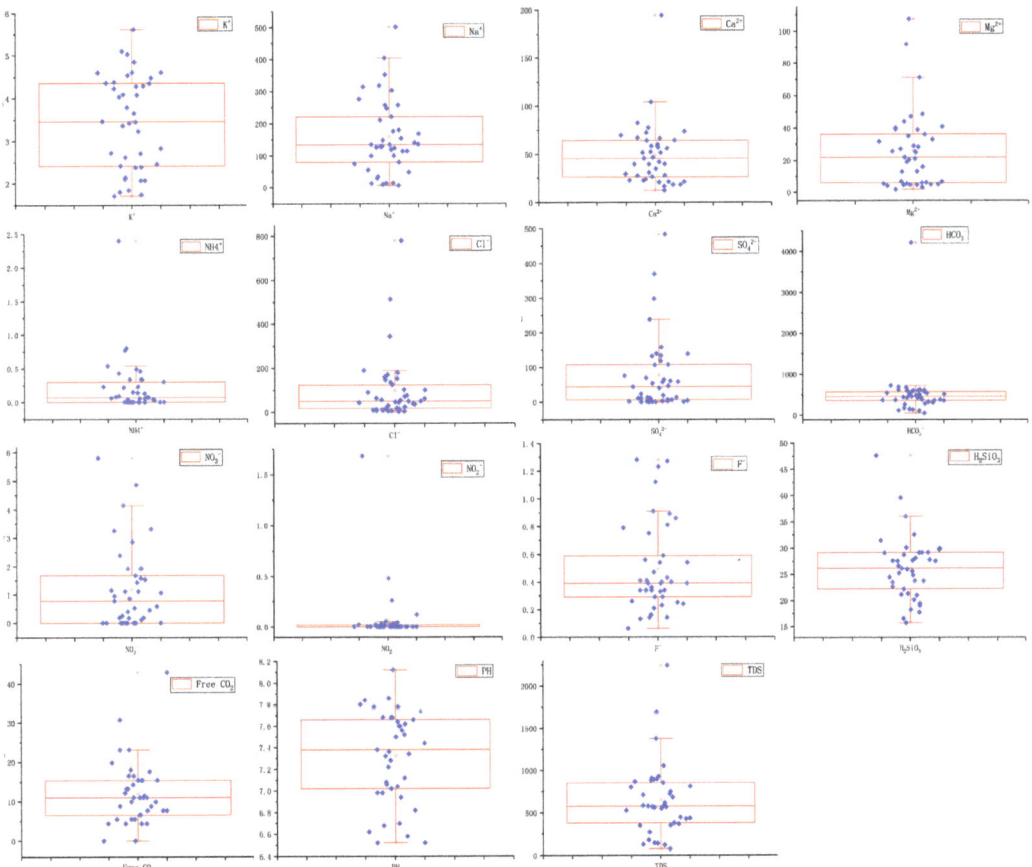

Figure 3. Boxplot of anion and anion in 41 samples.

For the above individual limit values, we can consider that they are caused by accidental factors, but the two measuring points CQ20 and CQ22 show limit values on multiple ion concentrations. From the map, it can be found that CQ20 and CQ22 are the two closest points to the Datang Coal Chemical Plant, which can confirm that the coal chemical plant has a certain impact on the groundwater in the region. The most severely exceeded ions in all samples in the region are NH_4^+, and NO_2^-. It shows that the local groundwater is polluted to a certain extent, and that the scope of influence of the coal chemical plant on the groundwater is extremely limited. Further analysis of the measuring points around the Dongming mining area shows that only DM4 exceeds the standard value in 14 points where NH_4^+ exceeds the standard value, and the concentration is less than 0.77 mg/L. Only DM6, DM12D, DM9 and DM11 exceed the standard value in 12 points where Na^+

exceeds the standard value, and the concentration is less than 314.42 mg/L, which shows that the impact of coal mining on the groundwater is also extremely limited. At least, it does not occupy a dominant position. So it can be analyzed that the local groundwater is mainly polluted by a large quantity of local cattle, sheep and other livestock manure, as well as industrial and domestic sewage pollution. On the other hand, there is a large amount of cultivated land distributed in the study area. In order to maintain the growth of crops, a large number of chemical fertilizers need to be used for a long time. The absorption capacity of crops to nitrogen fertilizer is weak, which increases the amount of nitrate in groundwater. Under the action of anaerobic microorganisms, nitrate in groundwater is reduced to nitrite and ammonia, which can increase the mass concentration of NH_4^+ and NO_2^-.

The Piper triple plot is a graphical representation of the concentration of major anions and cations in water, and is widely used to evaluate the relationship between dissolved ion components and major water types [41–43]. The three-line graph consists of three parts. The lower left corner and the lower right corner are two isosceles triangles representing the concentration of cations and anions, respectively. There is a diamond in the upper middle, and the side length of all graphs is 100 equal fractions. All data need to be normalized when plotting, representing the relative concentration of chemical components. Therefore, the ion concentration of each triangle will finally converge to a point. Then, the left and right triangles are used to make rays along the outer side parallel to the triangle, intersecting at a point within the diamond region. This point can represent the general chemical properties of groundwater and the relative composition of groundwater with anion cation pairs. The diamond was divided into nine zones. The characteristics of groundwater represented by each zone are shown in Figures 4 and 5.

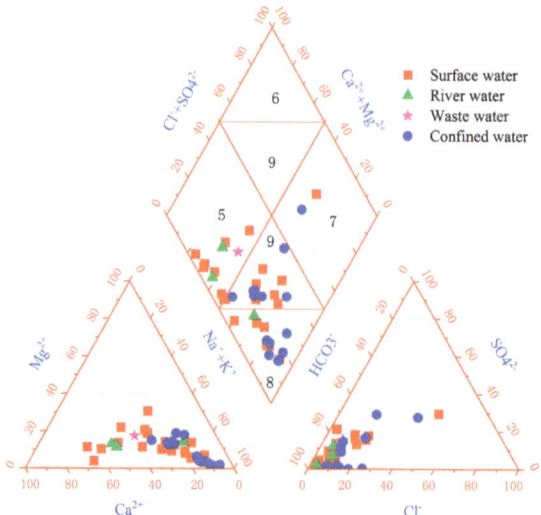

Figure 4. Comparison of watersamples of each aquifer Piper diagram.

We learn that the water chemical types in the study area are mainly HCO_3^- Na and HCO_3^- Ca·Na, the groundwater chemical types are mainly HCO_3^- Na and HCO_3^- Na·Ca, the confined water chemical type is HCO_3^- Na. In total, high salinity is low. Except for the two points of CQ20 and CQ22, all the points are in zone 3, indicating that the water samples in the study area show overall carbonation. All of the remaining confined water is in zones 8 and 9, showing Carbonate alkali (primary alkalinity) exceeding 50% and no cation-anion pairs exceeding 50%.

Sub-division	Characteristics/Nature of water
1	Alkaline earth (Ca+Mg) exceed alkalies (Na+K)
2	Alkalies exceed alkaline earths
3	Weak acids (CO_3 + HCO_3) exceed strong acids (SO_4 + Cl)
4	Strong acids exceed weak acid
5	Carbonate hardness (secondary alkalinity) exceed 50%
6	Non-carbonate hardness (secondary salinity) exceed 50%
7	Non-carbonate alkali (primary salinity) exceed 50%
8	Carbonate alkali (primary alkalinity) exceed 50%
9	No one cation-anion pairs exceed 50%

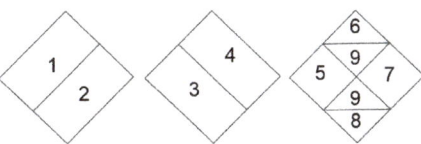

Figure 5. Piper Chart Division Water Quality Characteristics Interpretation Chart.

3.3. Influencing Factors of Hydrochemical Characteristics

3.3.1. Rock Leaching and Evaporation Concentration

The Gibbs chart is used to draw the correlation between the semi-log value of TDS and the ratio of $Na^+/(Na^+ + Ca^{2+})$ and $Cl^-/(Cl^- + HCO_3^-)$. Gibbs studied the major surface rivers in the world and divided the control factors of natural water into three types: evaporation concentration, rock weathering and atmospheric deposition [44].

As shown from Figures 6 and 7, most of the phreatic water and confined water in the study area fell in the middle of the figure, indicating that they were mainly affected by rock leaching and evaporation.

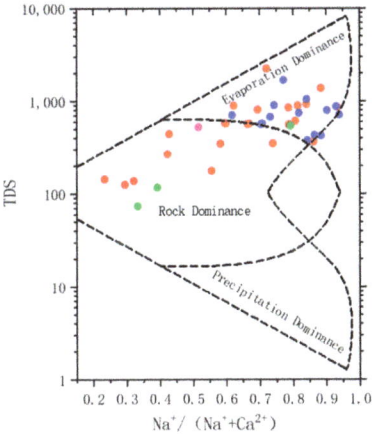

Figure 6. Gibbs distribution of cations in groundwater samples.

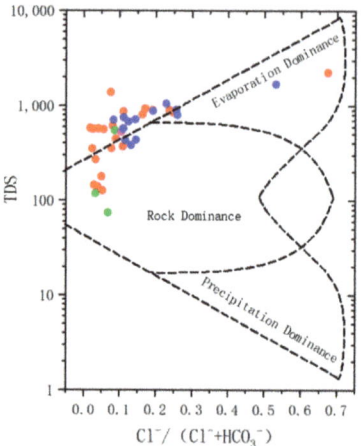

Figure 7. Gibbs distribution of anions in groundwater samples.

Similarly, the correlation between Mg^{2+}/Ca^{2+} and Mg^{2+}/Na^+ can be used to determine whether groundwater is affected by evaporation and salt leaching processes. As can be seen from Figure 8, the values of Mg^{2+}/Ca^{2+} and Mg^{2+}/Na^+ are relatively small. Except for a few two points, the values of other water sample points are all less than 1. Therefore, it can be considered that the groundwater in the study area is mainly affected by the dissolution of sodium salt and calcium salt.

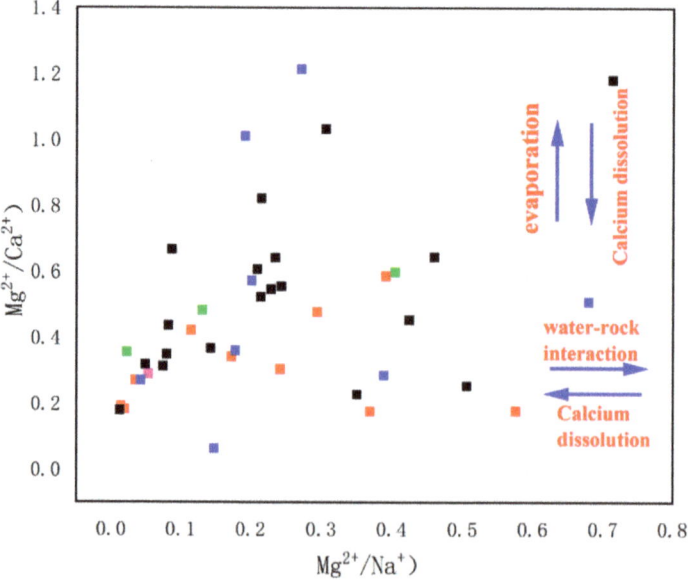

Figure 8. Distribution Map of Groundwater Ion Reaction.

3.3.2. Ion Exchange

In general, cation exchange can change some major chemical ions in groundwater [36]. Under natural conditions, the amount of Na^+ plus should be equal to the amount of Cl^- minus. As can be seen from the scatter diagram of Na^+ and Cl^- correlation in the study area (Figure 9), the water sample points of phreatic water and confined water in the study area are all below y = X, that is, the amount of Na^+ is greater than the amount of Cl^-, indicating that the excess Na^+ is the result of cation exchange.

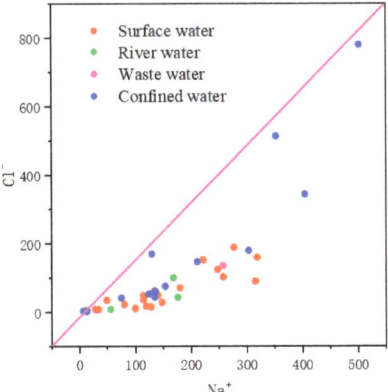

Figure 9. The correlation between Na^+ and Cl^-.

The values of $(Ca^{2+}+Mg^{2+}-HCO_3^--SO_4^{2-})/(Na^+-Cl^-)$ are often used to investigate whether Na^+ in aqueous media undergoes ion-exchange interactions with Ca^{2+} and Mg^{2+} in groundwater [21]. It can be seen from Figure 10 that $(Ca^{2+}+Mg^{2+}-HCO_3^--SO_4^{2-})$ is negatively correlated with (Na^+-Cl^-), and both are located on the y = −x line and below the point (0, 0), fully indicating that the excess Na^+ is obtained through the exchange of Na^+ in aqueous media with Ca^{2+} and Mg^{2+} in groundwater.

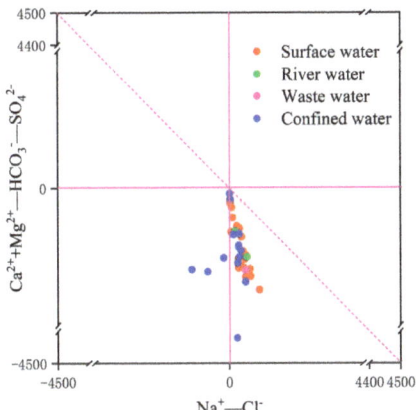

Figure 10. The correlation between $(Ca^{2+}+Mg^{2+}-HCO_3^--SO_4^{2-})$ and Na^+-Cl^-.

3.4. Grade Assessment of Groundwater Irrigation

Chenbaerhu Banner coalfield is located in the eastern grassland area, surrounded by pastoral areas, among which the Hulunbeier Grassland is one of the world-famous grasslands. It is worth discussing whether groundwater is suitable for irrigation. The percentage of Na and SAR in groundwater samples can affect the replacement of cations in clay minerals in soil, thus producing sodium or damaging soil permeability [45]. EC can reflect TDS concentration in groundwater [46,47]. Therefore, the Wilcox chart [48], with % Na, SAR and EC values as the classification standard, and the famous American Salinity

Laboratory Classification Chart [49], are used to evaluate the grade of irrigation water, in which the Wilcox chart represents both % Na and EC percentages, and the water samples are divided into five categories, namely C1S1, C2S1, C2S2, C3S1 and C4S1, which represent excellent, good permit, permit suspect, suspect unsuitable and unsuitable respectively. The calculation formulae are shown in Formulas (1) and (2), the calculation results are shown in Table 1, and the classification results are shown in Figures 11 and 12.

$$SAR = \frac{Na^+}{\sqrt{(Ca^{2+} + Mg^{2+})/2}} \quad (1)$$

$$\%Na^+ = \frac{K^+ + Na^+}{K^+ + Na^+ + Ca^{2+} + Mg^{2+}} \times 100\% \quad (2)$$

Table 1. Calculation results of %Na, SAR and EC.

Water Sample Number	SAR	EC	%Na
CQ1	33.79	579.77	0.72
CQ2	13.71	390.92	0.50
CQ3	0.72	386.41	0.60
CQ4	17.16	481.59	0.53
CQ5D	41.38	290.47	0.86
CQ6U	65.92	544.35	0.89
CQ7	2.22	98.26	0.24
CQ8	7.37	120.75	0.51
CQ9D	84.77	598.27	0.92
CQ10	3.17	94.34	0.32
CQ11	13.56	395.03	0.49
CQ12U	76.97	488.41	0.92
CQ13	27.13	381.29	0.72
CQ14U	30.87	258.38	0.81
CQ15	3.02	85.78	0.31
CQ16	26.98	508.52	0.67
DM10	6.93	302.34	0.35
DM11	36.00	630.75	0.73
CQ17	16.99	239.05	0.65
CQ18	18.81	388.72	0.61
CQ19	33.62	250.13	0.84
DM4	6.18	181.97	0.38
DM6	86.46	590.15	0.92
DM7	22.17	550.70	0.60
DM5	33.45	417.29	0.76
DM8	9.73	237.54	0.47
DM9	40.83	609.40	0.77
CQ20	34.20	1150.38	0.63
CQ21D	27.55	615.94	0.65
CQ22	43.49	1523.43	0.66
CQ23U	61.02	935.52	0.82
CQ24	13.99	605.84	0.44
CQ25D	36.38	294.98	0.83
DM12D	47.26	715.57	0.79
DM1	22.58	461.30	0.63
DM2	19.50	384.87	0.61
DM3	20.06	391.51	0.62
R1	23.69	372.37	0.68
R2	3.73	79.93	0.38
R3	2.19	50.36	0.35
W1	10.47	357.94	0.44

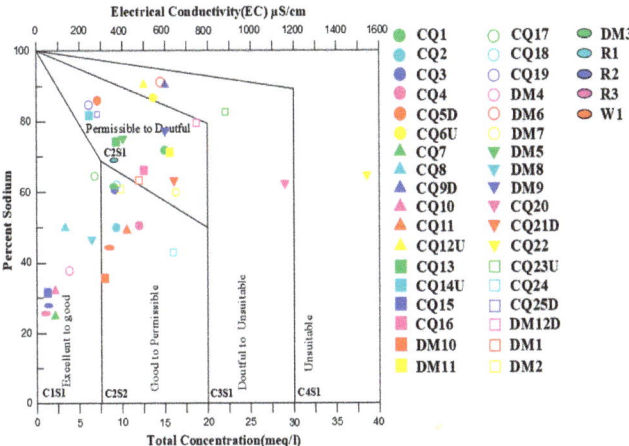

Figure 11. Wilcox diagramam.

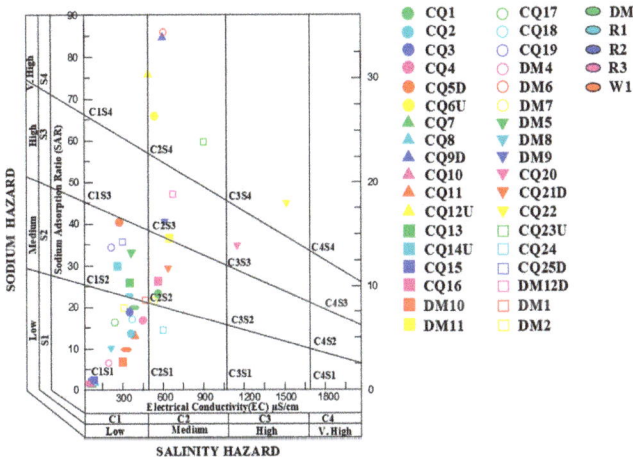

Figure 12. US Salinity Laboratory classification diagramam.

The USSL diagram shows that there are 19 water samples with high salinity and low alkalinity, namely CQ2, CQ3, CQ4, CQ7, CQ8, CQ10, CQ11, CQ15, DM10, CQ17, CQ18, DM4, DM8, CQ24, DM2, R3, R2, W1, and DM3; there were 11 water samples belonging to high salinity, moderate alkalinity, which were labeled CQ1, CQ5D, CQ13, CQ14U, CQ16, CQ19, DM5, CQ21D, CQ25D, DM1 and R1; there were 4 water samples with high salinity and high alkalinity, which were labeled DM11, DM9, CQ20 and DM12D; and the other 6 water samples are of very high salinity and high alkalinity, namely CQ6U, CQ12U, CQ9D, DM6, CQ23U and CQ22. The results of USSL map show that the local area as a whole is characterized by high salinity.

The results of Wilcox diagram show that 10 water samples from the Chenbaerhu Banner coalfield in Inner Mongolia are in the good permission range, indicating that they can be used as irrigation water. These water samples are numbered as CQ2, CQ3, CQ4, CQ11, DM10, CQ18, CQ24, DM2, R2 and W1. The water samples numbered CQ7, CQ8, CQ10, CQ15, CQ17, DM4, DM8, R2 and R3 were displayed in an excellent range, indicating that these nine water sample collection areas could be irrigated. Fifteen of the water samples were in the permit suspicious range. These were numbered CQ1, CQ5D,

CQ13, CQ14U, CQ16, DM11, CQ19, DM7, DM5, DM9, DM21D, CQ25D, DN12D, DM1, R1. The remaining six water samples are in suspect–unsuitable range, namely CQ6U, CQ9D, CQ12U, DM6, CQ20 and CQ23U. There was one water sample in the unsuitable interval, CQ22, indicating that the water sample collection area was not suitable for irrigation. In the analysis of seven samples in the suspicious–unsuitable section and unsuitable irrigation section, there are four samples belong to the confined water, accounting for 30.76% of the entire thirteen confined water samples, and there are three samples belongs to phreatic water, accounting for 12.5% of the entire thirteen confined water samples. For in-depth analysis of the external affected conditions of these seven points, CQ6U is located in the northwest of Chenbarhu Banner, and CQ20 is located in the northwest of Xiyutala town. Both are mainly affected by domestic water. CQ9D and CQ12U are located in the arid grassland at the foot of the Sanqi Mountain in the north of Chenbahu Banner. They are mainly affected by pasture production activities. CQ22 and CQ23U are located in farmland, and they are mainly affected by agricultural production activities. DM6 is located on the southeast side outside of the Dongming coal mine, separated with a road from the shortest mine boundary distance of 210 m, while DM7, only 10 m away from DM6, is in the permit suspicious range. In addition, a total of six samples were tested around the Dongming coal mine, except for DM6 and DM7, and the other samples are in the irrigable range, which indicates that the coal mine's impact on groundwater chemistry characteristics is extremely limited. DM6 and DM7 are more likely to be affected by the dust of road coal trucks, and the local groundwater pollution is mainly caused by a large quantity of local livestock manure such as that of cattle and sheep, industrial and domestic sewage pollution and cultivated land fertilizer.

4. Conclusions

(a) The main chemical types of river water are HCO_3^- Na and HCO_3^- Ca·Na, the main chemical types of surface water are HCO_3^- Na and HCO_3^- Na·Ca, and the main chemical type of confined water is HCO_3^- Na. The salinity of water samples in the study area is generally low, showing bicarbonate;

(b) Among the human factors affecting the change of hydrochemical characteristics, coal chemical plants and coal mining enterprises have limited influence on the change of groundwater chemical characteristics. The pollution of a large quantity of local cattle and sheep manure, industrial and domestic sewage and farmland fertilization are the main reasons for the increase of local underground NH_4^+ and NO_2^- mass concentrations, because the nitrate of groundwater is reduced to nitrite and ammonia under the action of anaerobic microorganisms;

(c) The natural factors for the change of hydrochemical characteristics mainly include the internal influence of rocks and water, which is mainly manifested in the leaching of sodium and calcium salts in rocks, and the ion exchange between Na^+ in aqueous media and Ca^{2+} and Mg^{2+} in groundwater;

(d) The results show that 46% of the water samples can be directly used for irrigation and 16% cannot be used for irrigation. Different water samples with different hydrochemical characteristics can be classified, and different measures can be taken to improve the use and management of groundwater chemistry characteristics. Regulating agricultural activities and sewage discharge is the main way to improve the water chemistry characteristics in the study area, to strengthen the monitoring of the groundwater environment, and to increase the investment in water source treatment.

Author Contributions: Conceptualization, W.S. and F.F.; methodology, W.S., F.F. and K.Y.; formal analysis, W.S., F.F. and Y.Z.; investigation, W.S., F.F. and K.Y.; data curation, W.S. and F.F.; writing—original draft preparation, W.S.; project administration, J.Z.; funding acquisition, W.S. and F.F.; revision and proof, J.S. All authors have read and agreed to the published version of the manuscript.

Funding: This work is supported by National Natural Science Foundation of China (No. 52104115); the Institute of Energy, Hefei Comprehensive National Science Center under Grant (No. 21KZS215); Independent project of the State Key Laboratory of Coal Mining Response and Disaster Prevention and Control (No. SKLMRDPC19ZZ08); Research on key technologies for development and utilization of abandoned mine resources and underground space (No. 20191101016).

Institutional Review Board Statement: Not applicable.

Informed Consent Statement: Not applicable.

Data Availability Statement: The authors confirm that the data supporting the findings of this study are available within the article.

Acknowledgments: We appreciate the support from the State Key Laboratory of Mining Response and Disaster Prevention and Control in Deep Coal Mines, and the Institute of Energy, Hefei Comprehensive National Science Center. Compliance with ethical standards.

Conflicts of Interest: The authors declare no conflict of interest.

References

1. Şehnaz, Ş.; Erhan, Ş.; Ayşen, D. Evaluation of water quality using water quality index (WQI) method and GIS in Aksu River (SW-Turkey). *Sci. Total Environ.* **2017**, *584–585*, 131–144.
2. Adimalla, N.; Li, P.; Venkatayogi, S. Hydrogeochemical evaluation of groundwater quality for drinking and irrigation purposes and integrated interpretation with water quality index studies. *Environ. Process.* **2018**, *5*, 363–383. [CrossRef]
3. Selemani, J.R.; Zhang, J.; Muzuka, A.N.; Njau, K.N.; Zhang, G.; Maggid, A.; Mzuza, M.K.; Jin, J.; Pradhan, S. Seasonal water chemistry variability in the Pangani River basin, Tanzania. *Environ. Sci. Pollut. Res.* **2017**, *24*, 26092–26110. [CrossRef] [PubMed]
4. Varol, S.; Davraz, A. Evaluation of the groundwater quality with WQI (Water Quality Index) and multivariate analysis: A case study of the Tefenni plain (Burdur/Turkey). *Environ. Earth Sci.* **2015**, *73*, 1725–1744. [CrossRef]
5. Edet, A.; Ukpong, A.; Nganje, T. Hydrochemical studies of Cross River Basin (southeastern Nigeria) river systems using cross plots, statistics and water quality index. *Environ. Earth Sci.* **2013**, *70*, 3043–3056. [CrossRef]
6. Arumugam, K.; Elangovan, K. Hydrochemical characteristics and groundwater quality assessment in Tirupur region, Coimbatore district, Tamil Nadu, India. *Environ. Geol.* **2009**, *58*, 1509. [CrossRef]
7. Chenini, I.; Khmiri, S. Evaluation of ground water quality using multiple linear regression and structural equation modeling. *Int. J. Environ. Sci. Technol.* **2009**, *6*, 509–519. [CrossRef]
8. Galbraith, L.M.; Burns, C.W. Linking land-use, water body type and water quality in southern New Zealand. *Landsc. Ecol.* **2007**, *22*, 231–241. [CrossRef]
9. Rao, Y.S.; Reddy, T.; Nayudu, P. Groundwater quality in the Niva river basin, Chittoor district, Andhra Pradesh, India. *Environ. Geol.* **1997**, *32*, 56–63.
10. Fauriel, S.; Laloui, L. A bio-chemo-hydro-mechanical model for microbially induced calcite precipitation in soils. *Comput. Geotech.* **2012**, *46*, 104–120. [CrossRef]
11. Choo, J.; Sun, W. Cracking and damage from crystallization in pores. Coupled chemo-hydro-mechanics and phase field modeling. *Comput. Methods Appl. Mech. Eng.* **2018**, *335*, 347–379. [CrossRef]
12. Ghorbani, J.; El-Zein, A.; Airey, D.W. Thermo-elasto-plastic analysis of geosynthetic clay liners exposed to thermal dehydration. *Environ. Geotech.* **2018**, *8*, 566–580. [CrossRef]
13. Penga, S.; Fenga, F.; Dua, W.; Hea, Y.; Chonga, S.; Xinga, Z. Analysis of water chemical characteristics and application around large opencast coal mines in grassland: A case study of the North Power Shengli coal mine. *Desalination Water Treat.* **2019**, *141*, 149–162. [CrossRef]
14. Tang, K.; Hou, J.; Tang, K. Assessment of groundwater quality in China: I. Hydrochemical characteristics of groundwater in plain area. *Water Resour. Prot.* **2006**, *22*, 1–5.
15. He, X.; Wu, J.; He, S. Hydrochemical characteristics and quality evaluation of groundwater in terms of health risks in Luohe aquifer in Wuqi County of the Chinese Loess Plateau, northwest China. *Hum. Ecol. Risk Assess. Int. J.* **2019**, *25*, 32–51. [CrossRef]
16. Cao, G.Y.; Yang, H.T.; Ren, Y.J. Hydrogeochemical characteristics and causes of groundwater in Chengjiaying Basin, Inner Mongolia. *Geol. Chem. Miner.* **2019**, *41*, 285–290.
17. Xiao, M.; Han, Z.; Xu, S.; Wang, Z. Temporal Variations of Water Chemistry in the Wet Season in a Typical Urban Karst Groundwater System in Southwest China. *Int. J. Environ. Res. Public Health* **2020**, *17*, 2520. [CrossRef]
18. Howladar, M.F.; Deb, P.K.; Muzemder, A.S.H.; Ahmed, M. Evaluation of water resources around Barapukuria coal mine industrial area, Dinajpur, Bangladesh. *Appl. Water Sci.* **2014**, *4*, 203–222. [CrossRef]
19. JKim, G.; Ko, K.-S.; Kim, T.H.; Lee, G.H.; Song, Y.; Chon, C.-M.; Lee, J.-S. Effect of mining and geology on the chemistry of stream water and sediment in a small watershed. *Geosci. J.* **2007**, *11*, 175–183.
20. Lin, D.; Qiang, W.; Zhang, R.; Song, Y.; Chen, S.; Pei, L.; Liu, S.; Bi, C.; Lv, Z.; Huang, S. Environmental characteristics of groundwater: An application of PCA to water chemistry analysis in Yulin. *J. China Univ. Min. Technol.* **2007**, *17*, 73–77.

21. Zhang, B.; Zhao, D.; Zhou, P.; Qu, S.; Liao, F.; Wang, G. Hydrochemical Characteristics of Groundwater and Dominant Water–Rock Interactions in the Delingha Area, Qaidam Basin, Northwest China. *Water* **2020**, *12*, 836. [CrossRef]
22. Cario, G.; Casavola, A.; Gjanci, P.; Lupia, M.; Petrioli, C.; Spaccini, D. Long lasting underwater wireless sensors network for water quality monitoring in fish farms. In Proceedings of the OCEANS 2017-Aberdeen, Aberdeen, UK, 19–22 June 2017; pp. 1–6.
23. Mahmud, M.A.; Hussain, K.A.; Hassan, M.; Jewel, A.R.; Shamsad, S.Z. Water quality assessment using physiochemical parameters and heavy metal concentrations of circular rivers in and around Dhaka city, Bangladesh. *Int. J. Water Res.* **2017**, *7*, 23–29.
24. Hou, H.; Zhou, J.; Liu, G.; Kong, J.; Enyao, M.; Luo, W. Study on the geographical origin identification of american ginseng based on multi-element analysis and statistical methods. *Hubei Agric. Sci.* **2020**, *59*, 151–154.
25. Arkoc, O.; Ucar, S.; Ozcan, C. Assessment of impact of coal mining on ground and surface waters in Tozaklı coal field, Kırklareli, northeast of Thrace, Turkey. *Environ. Earth Sci.* **2016**, *75*, 514. [CrossRef]
26. Zhang, H.; Su, L.; Wang, J.; Yang, L.; Wang, D.; Hu, X.; Xiong, L. Study on LA-ICP-MS Determination of Trace Elements in Sulfide Minerals. *Hans J. Chem. Eng. Technol.* **2019**, *9*, 401–409. [CrossRef]
27. Jiang, Q.; Han, Y.; Sun, X.; Gong, H.; Qian, W.; Guoxing, L.U. Study on the Determination and Its Difference Analysis of Chloride and Sulfate in Different Soils by Ion Chromatography and Capillary Electrophoresis. *Soils* **2016**, *48*, 343–348.
28. Durowoju, O.S.; Ekosse GI, E.; Odiyo, J.O. Occurrence and Health-Risk Assessment of Trace Metals in Geothermal Springs within Soutpansberg, Limpopo Province, South Africa. *Int. J. Environ. Res. Public Health* **2020**, *17*, 4438. [CrossRef]
29. Lai, Z.; Lin, F.; Qiu, L.; Wang, Y.; Chen, X.; Hu, H. Development of a sequential injection analysis device and its application for the determination of Mn(II) in water. *Talanta* **2020**, *211*, 120752. [CrossRef]
30. General Administration of Quality Supervision I.a.Q.; China S.A.o. *Standard for Groundwater Quality*; China Environmental Science Press: Beijing, China, 2017; p. 20.
31. Choi, H.; Poythress, J.C.; Park, C.; Jeon, J.J.; Park, C. Regularized boxplot via convex clustering. *J. Stat. Comput. Simul.* **2019**, *89*, 1227–1247. [CrossRef]
32. Khare, P. A large-scale investigation of the quality of groundwater in six major districts of Central India during the 2010–2011 sampling campaign. *Environ. Monit. Assess.* **2017**, *189*, 429. [CrossRef] [PubMed]
33. Singh, K.K.; Tewari, G.; Kumar, S. Evaluation of Groundwater Quality for Suitability of Irrigation Purposes: A Case Study in the Udham Singh Nagar, Uttarakhand. *J. Chem.* **2020**, *2020*, 6924026. [CrossRef]
34. Kumar, P.S.; Balamurugan, P. Evaluation of groundwater quality for irrigation purpose in Attur taluk, Salem, Tamilnadu, India. *Water Energy Int.* **2018**, *61*, 59–64.
35. Haile, E.; Fryar, A.E. Chemical evolution of groundwater in the Wilcox aquifer of the northern Gulf Coastal Plain, USA. *Hydrogeol. J.* **2017**, *25*, 2403–2418. [CrossRef]
36. Davis, A.; Heatwole, K.; Greer, B.; Ditmars, R.; Clarke, R. Discriminating between background and mine-impacted groundwater at the Phoenix mine, Nevada USA. *Appl. Geochem.* **2010**, *25*, 400–417. [CrossRef]
37. Sefati, Z.; Khalilimoghadam, B.; Nadian, H. Assessing urban soil quality by improving the method for soil environmental quality evaluation in a saline groundwater area of Iran. *Catena* **2019**, *173*, 471–480. [CrossRef]
38. Shakerkhatibi, M.; Mosaferi, M.; Pourakbar, M.; Ahmadnejad, M.; Safavi, N.; Banitorab, F. Comprehensive investigation of groundwater quality in the north-west of Iran: Physicochemical and heavy metal analysis. *Groundw. Sustain. Dev.* **2019**, *8*, 156–168. [CrossRef]
39. Wang, X.; Xiao, W.; Liu, H. Soil moisture characteristic curve and prediction of available water content of overburden in Xilinhot Mining Area. *Coal Sci. Technol.* **2020**, *48*, 169–177.
40. Liu, H.; Huang, H.; Shuai, B.; Feng, Y. Study on Stability of East Side Slope of Shengli East No. 2 Mine Based on Geo-Studio Numerical Software. *Adv. Geosci.* **2020**, *10*, 622–628. [CrossRef]
41. Alhamed, M. The hydrological and the hydrogeological framework of the Lottenbachtal, Bochum, Germany. *Appl. Water Sci.* **2017**, *7*, 315–328. [CrossRef]
42. Arslan, B.; Akün, E. Management, contamination and quality evaluation of groundwater in North Cyprus. *Agric. Water Manag.* **2019**, *222*, 1–11. [CrossRef]
43. Benmoussa, Y.; Remini, B.; Remaoun, M. Quality assessment and hydrogeochemical characteristics of groundwater in Kerzaz and Beni Abbes along Saoura valley, southwest of Algeria. *Appl. Water Sci.* **2020**, *10*, 170. [CrossRef]
44. Gibbs, R.J. Mechanisms controlling world water chemistry. *Science* **1970**, *170*, 1088–1090. [CrossRef] [PubMed]
45. Liu, J.; Jin, D.; Wang, T.; Gao, M.; Yang, J.; Wang, Q. Hydrogeochemical processes and quality assessment of shallow groundwater in Chenqi coalfield, Inner Mongolia, China. *Environ. Earth Sci.* **2019**, *78*, 347. [CrossRef]
46. Mahato, M.K.; Singh, P.K.; Tiwari, A.K. Hydrogeochemical evaluation of groundwater quality and seasonal variation in East Bokaro coalfield region, Jharkhand. *J. Geol. Soc. India* **2016**, *88*, 173–184. [CrossRef]
47. Mahato, M.K.; Singh, P.K.; Singh, A.K.; Tiwari, A.K. Assessment of hydrogeochemical processes and mine water suitability for domestic, irrigation, and industrial purposes in East Bokaro Coalfield, India. *Mine Water Environ.* **2018**, *37*, 493–504. [CrossRef]
48. Wilcox, L. *Classification and Use of Irrigation Waters*; US Department of Agriculture: Washington, DC, USA, 1955.
49. Handa, B. Studies on US Salinity Laboratory Diagram for Classification of Irrigation Waters. *J. Indian Soc. Soil Sci.* **1965**, *13*, 227–232.

Article

Effect of Continuous Loading Coupled with Wet–Dry Cycles on Strength Deterioration of Concrete

Linzhi Wang [1,2], Mingzhong Gao [1,2,*] and Jiqiang Zhang [1,2]

1. State Key Laboratory of Mining Response and Disaster Prevention and Control in Deep Coal Mines, Anhui University of Science and Technology, Huainan 232001, China
2. National & Local Joint Engineering Research Center of Precision Coal Mining, Anhui University of Science and Technology, Huainan 232001, China
* Correspondence: mzgao@aust.edu.cn; Tel.: +86-180-5542-7798

Abstract: In practical engineering, concrete is often under continuous stress conditions and there are limitations in considering the effect of wet–dry cycles alone on the strength deterioration of concrete. In order to study the deterioration of concrete strength under the coupling of load and wet-dry cycles, concrete specimens were loaded with 0%, 10%, 20%, and 35% stress levels and coupled to undergo one, three, and seven wet–dry cycles. The strength deterioration of the concrete was obtained by uniaxial compression and the regression equation was established. The strength deterioration mechanism of the concrete under the coupled conditions was analyzed and revealed through an AE acoustic emission technique and nuclear magnetic resonance technique. The results of the study show that, with the same number of wet–dry cycles, there are two thresholds of a and b for the uniaxial compressive strength of concrete with the stress level, and with the progression of wet–dry cycles, the length of the interval from a to b gradually shortens until it reaches 0. The cumulative AE energy of concrete decreases with the progression of wet–dry cycles; using the initiating crack stress as the threshold, the calm phase of concrete acoustic emission, the fluctuating phase, and the NMR T_2 spectral peak area show different patterns of variation with the increase in the number of wet–dry cycles.

Keywords: concrete; sustained compressive loading; wet–dry cycles; damage evolution; regression analysis; acoustic emission; nuclear magnetic resonance

1. Introduction

In engineering applications, the study of the mechanical properties of concrete structures under natural conditions plays a vital role in effectively predicting the life cycle of buildings. The factors affecting the mechanical properties of concrete include carbonation [1–3], freeze–thaw cycles [4–6], wet–dry cycles (chloride salts [7–9], as well as sulphate [10,11]), and the effects of different forms of loading [12–16], where the effects of wet–dry cycles of different salt solutions on the mechanical properties of concrete have become the focus of a wide range of scholars.

As concrete is made of rocks that experience natural effects, making its chemical composition not yet stable, an in-depth study of the role of different salt solutions for wet–dry cycles on the impact of the concrete structure is of great significance. For example, Su [17] argued that the mass fraction of the salt solution is directly proportional to the degree of concrete damage, and with the progression of wet–dry cycles, the quality of concrete first increased and then decreased. Wang [18] used SEM technology to explain the mechanism of concrete deterioration due to the crystalline swelling effect of $Na_2SO_4 \cdot 10H_2O$ from a microscopic perspective, followed by Jiang [19], through regression analysis, who established the wet–dry cycle conditions of a sulphate solution using a stress–strain equation for damaged concrete under wet–dry cycles of a sulphate solution. Although there are many saline areas

in China, the concentration of salt solutions subjected to wet–dry cycles in most areas of buildings is very low or even negligible, and it is equally important to study the effect of water on the wet–dry cycles of concrete. However, in practical engineering, concrete is often in distinctive low stress situations, and the single factor consideration of wet–dry cycling makes the prediction of concrete durability somewhat limited, and it is not conducive to accurately deriving the service life of buildings. Combined with the current research, the summary of the coupling of wet–dry cycles and continuous loading on the concrete deterioration law is not extensive enough; Wang [20] completed a similar exploration of CFRP reinforced beams, and revealed the method of damage from the surface layer to the internal bond development; however, for concrete material, the coupling effect related to the deterioration mechanism has not yet been explored.

This paper uses a rock rheological perturbation effect instrument as a continuous load system, and utilizes natural immersion at room temperature and natural drying to simulate wet–dry cycles. The effect of different loads coupled with wet–dry cycles on the deterioration process of concrete strength is analyzed, and the deterioration mechanism of concrete is analyzed according to acoustic emission and nuclear magnetic resonance techniques. The results of the research can be invoked as a reference for the accurate prediction of the service life of concrete structures under combined dry, wet, and continuous loading, as well as for protective measures.

2. Materials and Methods

2.1. Material Composition

The concrete mixes are shown in Table 1. Also, the individual materials are described as follows.

Table 1. Mix proportion of concrete kg/m^3.

Material	Cement	Granite Gravel	Sand	Water
Quality/kg	404.04	808.08	808.08	202.02

Cement: M32.5 masonry cement (cement from Huainan Shunyue Cement Co., Ltd., Huainan, China) is used.

Granite gravel: Particle size 5–15 mm, density 2.63 g/cm^3.

Sand: Huaihe River sand, maximum particle size 4.75 mm, fineness modulus 2.4, density 2.41 g/cm^3.

Water: Laboratory tap water.

A PVC pipe with an outer diameter of 15 mm and a length of 200 mm was inserted into the bottom center of a standard concrete mold, as shown in Figure 1a. Cement was poured into this mold, then the PVC pipe was pulled out promptly within 6 h after fabrication. Finally, the molds were removed after 24 h and placed in saturated calcium hydroxide solution for 28 d. The details are shown in Figure 1b.

(a)   (b)

Figure 1. Concrete preparation, including the (**a**) mold and (**b**) concrete specimens.

Three standard cubic specimens (150 mm × 150 mm × 150 mm) were prepared for determining the compressive strength of concrete in standard curing for 28 d. The uniaxial compressive strength result was 17.7 MPa. Sixteen hollow concrete specimens were prepared, three of which were used to determine the compressive strength of the hollow concrete (17.7 MPa), one served as a control group, and 12 experienced the effects of different coupling conditions. In order to reduce the interference of the creep effect on the deterioration process, this test only discussed the first few wet—dry cycles.

2.2. Design of the Degradation Process by Coupled Load-Holding and Wet–Dry Cycles

In order to reduce the fluctuation of the continuous pressure load, this test adopted the mechanical loading method using the RRTS-II Rock Rheology and Disturbance Effect Tester [21,22] as the loading device. When loading, firstly, the hydraulic oil was delivered into the small cylinder and then the big cylinder through the hydraulic pump, then the piston rod in the big cylinder contacted the concrete and provided the compressive stress, and finally the small cylinder maintained the pressure in the big cylinder through the pipeline, thus constituting a pressure stabilization system, as shown in Figure 2. Loading was done using gears and hydraulic secondary expansion, with an expansion ratio of up to 60–100 times or so (the expansion ratio of this test device is 72); the expansion ratio K can be expressed as follows:

$$K = \frac{d_1}{d_2} \times \frac{\varphi_1^2}{4\varphi_2^2} \tag{1}$$

where d_1 and d_2 are the diameters of the large and small gears, respectively, and φ_1 and φ_2 are the diameters of the pistons of the large and small cylinders, respectively.

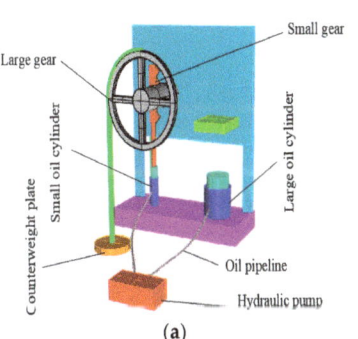 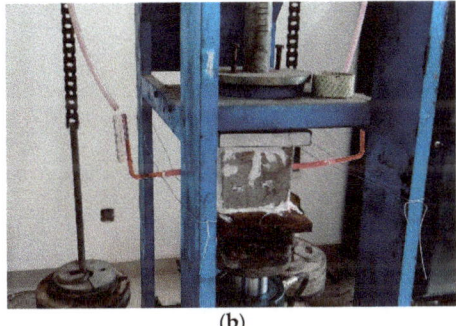

Figure 2. Testing machine used for the rock rheological perturbation effect, including the (**a**) schematic and (**b**) actual test.

Define the load level λ_c as the ratio of the actual loading stress f to the mean uniaxial compressive strength of the specimen f_c, which can be expressed as follows:

$$\lambda_c = \frac{f}{f_c} \times 100\% \tag{2}$$

The concrete water absorption system is based on the principle of the linker. Two rubber tubes (13 mm outer diameter) are glued to the two ends of the concrete central aperture, and the joints are filled with glass glue (left for 24 h) to ensure the sealing of the water absorption system. After 10 min of applying load to the concrete in order to reach a predetermined value, two L-shaped (6 mm inner diameter) glass tubes were inserted into the rubber tubes. Then, water was injected from the inlet pipe, and after 24 h of water absorption, the blower was used to dry naturally for 24 h. After 24 h of the water absorption process, the concrete surface showed obvious water stains, and the initial water absorption rate of concrete was approximated after 24 h and 48 h of air drying time, respectively.

Therefore, it is considered that the concrete reached the natural drying state after 24 h of air drying

The degradation coupling conditions of this test are shown in Table 2, and at the end of the degradation process, four standard cylindrical specimens with a diameter of 50 mm and a height of 100 mm were removed from each hollow concrete specimen. One of them was soaked in water until saturation for NMR testing, and the remaining three were dried in an oven at 106 °C for 48 h for uniaxial compressive strength testing, while the average moisture content was 2% in the natural drying state and 6% in the saturated state.

Table 2. Coupling conditions to which the concrete was subjected.

Number	0-1	0-3	0-7	1-1	1-3	1-7	2-1	2-3	2-7	3-1	3-3	3-7
Loading/MPa	0	0	0	1.77	1.77	1.77	3.54	3.54	3.54	6.20	6.20	6.20
Number of wet-dry cycles/N	1	3	7	1	3	7	1	3	7	1	3	7

As shown in Figures 3 and 4, Figure. 3 shows the concrete undergoing the process of coupled action of continuous loading and wet-dry cycles. Figure 4 shows the location of each concrete specimen coring ($\Phi 50$ mm × 100 mm), and the overall coring.

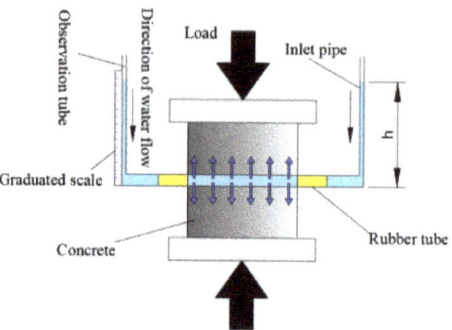

Figure 3. Concrete loading and water absorption processes.

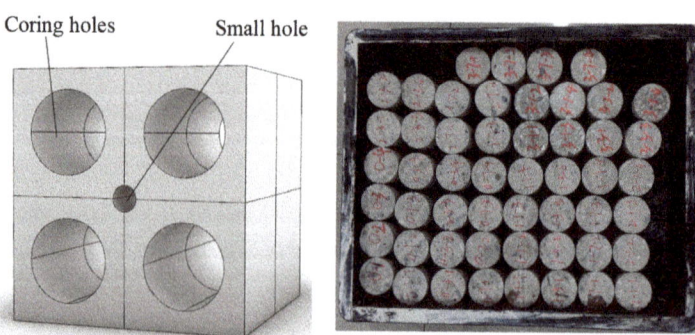

Figure 4. Concrete coring.

2.3. Nuclear Magnetic Resonance Test

One specimen was taken from each of the 12 groups with different coupling conditions and one control group, and then immersed in water for 48 h (the difference between the masses after 24 h and 48 h was less than 0.1 g; the specimens after 48 h of immersion in water were considered as saturated specimens). To prevent moisture dissipation, the specimens were removed from the water, immediately wrapped tightly with plastic wrap, and then put into the NMR analyzer one by one for the T_2 spectrum testing.

2.4. Uniaxial Compression and Acoustic Emission Tests

A uniaxial compression test with a loading rate of 0.01 mm/s was performed on the remaining three specimens of 12 groups with different coupling condition degradation groups and one control group. The AE system was also used for real-time monitoring during the loading process, with the acquisition threshold set to 40 dB and the acquisition frequency set to 3.5 MHz

3. Deterioration Law and Analysis of Concrete Strength Parameters
3.1. Strength Law after Concrete Deterioration

The uniaxial compressive strength and modulus of elasticity of the concrete under the influence of different deterioration conditions are shown in Table 3.

Table 3. Uniaxial compressive strength and modulus of elasticity of concrete after deterioration.

Number	Coupling Conditions		Actual Experimental Value σ_c/MPa	σ_c/MPa	E_c/GPa
	Stress Levels/λ_c	Number of Wet-Dry Cycles/N			
0	0%	0	18.14 17.30 16.40	17.30	4.33
0-1	0%	1	14.53 14.30 13.54	14.30	3.28
0-3	0%	3	14.33 13.70 12.86	13.70	2.58
0-7	0%	7	12.53 11.57 10.86	11.57	2.25
1-1	10%	1	13.89 13.50 12.94	13.50	3.61
1-3	10%	3	12.89 12.20 11.46	12.20	3.42
1-7	10%	7	11.16 10.39 10.12	10.39	2.27
2-1	20%	1	16.03 15.18 14.89	15.18	3.82
2-3	20%	3	15.47 14.88 14.36	14.88	3.02
2-7	20%	7	10.38 9.73 9.55	9.73	1.44
3-1	35%	1	14.67 13.50 13.43	13.50	3.35
3-3	35%	3	13.34 13.06 12.57	13.06	2.31
3-7	35%	7	9.36 8.45 8.33	8.45	1.18

To determine whether the effect of stress level and the number of wet–dry cycles on the uniaxial compressive strength of concrete is significant, based on Table 3, a two-factor ANOVA was performed using origin software, as shown in Table 4. It was found that the p-values of the stress level, the number of wet–dry cycles, and the interaction between the two factors were all less than 0.05. It can be concluded that both the stress level and number of wet–dry cycles had a significant effect on the uniaxial compressive strength of concrete.

Table 4. Results of two-way ANOVA.

	DF	SST	SD	F	P
Stress level	3	15.68	5.22	14.42	1.40×10^{-5}
Number of wet-dry cycle	2	107.88	53.94	148.82	2.98×10^{-14}
Interaction	6	15.97	2.66	7.34	1.53×10^{-4}
Model	11	139.53	12.68	34.99	4.44×10^{-12}
Errors	24	8.70	0.36	–	–

Based on the data in Table 3, the effects of the coupled condition on the deterioration of concrete are discussed in terms of the number of wet–dry cycles and stress levels, respectively, as shown in Figure 5

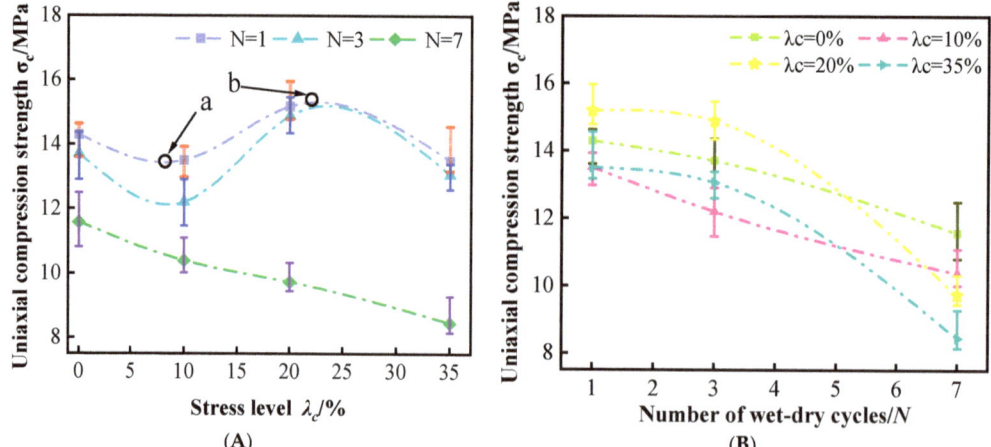

Figure 5. Variation of specimen σ_c under different deterioration conditions, including (**A**) different stress levels and (**B**) different number of wet–dry cycles.

As shown in Figure 5A, the effect of different stress levels on concrete uniaxial compressive strength deterioration also differed when the number of wet–dry cycles was the same. There were two thresholds of a and b for the effect of the stress level on the uniaxial strength deterioration of concrete in both the first and third wet–dry cycles, where the stress level was 0% for interval a and 35% for interval b, and the uniaxial compressive strength of the concrete was negatively correlated with the stress level. However, in the interval from a to b, the uniaxial compressive strength of concrete increased with the increase in stress level, and at the same time, the length of the interval in the range from a to b gradually decreased with the increase in the number of wet–dry cycles.

The analysis showed that, as the stress level increased, the pores inside the concrete first closed and then ruptured to develop cracks. When the stress level was in the interval of 0% to a, the pores started to close and the pore size kept shrinking. According to the literature [23,24], the height of capillary water absorption inside the concrete is inversely proportional to the capillary pore size. Therefore, as the stress level increases, the water invades deeper into the concrete, increasing the contact area between the particles and the water. Then, the bond between the particles is weakened, making the uniaxial compressive strength of concrete decreasing. In the stress level in the interval a to b, the pore closure, to a certain extent, made the pore diameter smaller than the capillary pore diameter, thus gradually blocking the water to the internal erosion. Therefore, in this interval, the uniaxial compressive strength of concrete was positively related to the stress level. When the stress level was at b to 35%, the pores inside the concrete gradually expanded and converged into cracks under the stress. This led to an increase in the contact area between the internal particles of the concrete and water, making its uniaxial compressive strength decrease with the increasing stress level. In addition, this analysis could be verified at different numbers of wet–dry cycles affecting the length of interval from a to b. After different times of wet–dry cycle deterioration, the uniaxial compressive strength of concrete decreased, resulting in a corresponding decrease in the initiating crack stress and a corresponding decrease in threshold b.

As shown in Figure 5B, the overall concrete uniaxial compressive strength decreased with the increase in the number of wet–dry cycles. However, the decreasing trend of uniaxial compressive strength varied under different stress levels. At a stress level λ_c of 0%, the uniaxial compressive strength of concrete decreased approximately linearly. Nevertheless, at a stress level λ_c of 10%, the uniaxial compressive strength of the concrete decreased in a concave curve. This is because the bond between the particles inside the concrete was weakened by the external load when a continuous load with a stress level λ_c of 10% was applied to the concrete. At this time s, the first few wet–dry cycles of concrete deterioration increased. However, with the increase in the number of wet–dry cycles, the bond between the internal particles of concrete was reduced. Thus, the concrete uniaxial compressive strength tended to stabilize. At stress levels λ_c of 20% and 35%, the uniaxial compressive strength of concrete showed a convex non-linear decreasing trend. This was because for the stress level at this stage, the concrete internal cracks started to develop, while water intrusion at the cracks dissolved the cement between the particles. At the same time, with the increase in the number of wet–dry cycles, the repeated dissolution of water on the particles at the concrete fissured. This led to the development of more cracks, which in turn accelerated the rate of concrete strength decline.

3.2. Regression Analysis of Uniaxial Compressive Strength of Concrete

To analyze the variation of the uniaxial compressive strength of concrete under the action of different coupling conditions we used non-linear surface fitting of the uniaxial compressive strength of concrete, with the stress level and number of wet–dry cycles as independent variables in the software origin. After several fitting comparisons, the uniaxial compressive strength RationalTaylor nonlinear surface regression model was obtained, as shown in Equation (3).

$$z = \frac{15.06 - 9.67x - 7.78y + 0.07y^2 + 2.71xy}{1 - 0.62x - 0.53y - 0.02x^2 + 0.2xy} \quad R^2 = 0.804 \tag{3}$$

where z is the uniaxial compressive strength of concrete after deterioration, x is the number of wet–dry cycles, and y is the stress level.

As shown in Figure 6, a visualization model was constructed to analyze the variation of the uniaxial compressive strength of concrete after the action of different coupling conditions. Overall, the relationship between the stress level and the uniaxial compressive strength of concrete tended to be gradually negative from the threshold fluctuations as the number of wet–dry cycles increased. Overall, with the increase in the number of wet–dry cycles, the relationship between the stress level and uniaxial compressive strength of concrete fluctuated from a threshold value then gradually tended to have a negative correlation.

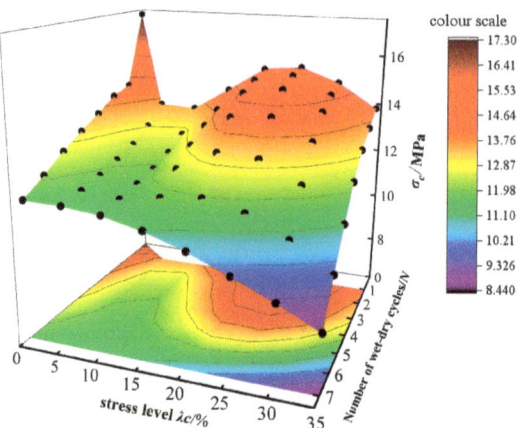

Figure 6. Strength deterioration pattern of the coupled concrete.

3.3. Analysis of Concrete Damage Evolution

Numerous studies have shown that the damage variable D of concrete under external loading should satisfy the Weibull distribution, whose expression is as follows:

$$D = 1 - exp[-(\zeta/a)^m] \tag{4}$$

where ζ is the strain, m is the shape parameter, and a is the material parameter. According to the summary of Wu [25], the larger the value of m, the more elastic or brittle the material tends to be, and the smaller the value of m, the more plastic the material tends to be.

Also according to the literature [25], parameters m and a are related to the material properties as follows:

$$E_0 = \frac{\sigma_{max}}{\varepsilon_{max}} \tag{5}$$

$$m = \frac{1}{\ln(E) - \ln(E_0)} \tag{6}$$

$$a = \frac{\zeta_{max}}{\left(\frac{1}{m}\right)^{(1/m)}} \tag{7}$$

where E is the initial modulus of elasticity of the concrete, E_0 is the cut-line modulus of the concrete past the peak load point after deterioration, and ζ_{max} is the strain corresponding to the maximum stress value of the concrete after deterioration.

To facilitate a comparison of the concrete damage curves under different deterioration conditions, define $\zeta/\zeta_{max} = x$ and substitute into Equation (4), i.e.,

$$D = 1 - exp[-(x/a)^m] \tag{8}$$

We then completed an analysis of the initial damage to the concrete after the action of different coupling conditions and the damage during uniaxial compression. For reasons of space, the damage curves for concrete subjected to seven wet–dry cycles and after a stress level of 35% deterioration were compared separately in this paper. The relevant parameters of the damaged specimens are shown in Table 5.

As shown in Figure 7a, the growth rate of the concrete damage variable D at the initial stage was proportional to the stress level, until the strain was less than the peak strain. For both, after the strain ratio was greater than 0.75, the damage variable D tended to level off. The degree of deterioration of concrete deepened with the increase in the stress level of the load applied after the action of multiple wet–dry cyclic processes.

Table 5. Selected concrete damage related parameters.

Number	σ_{max}/MPa	$\varepsilon_{max}/10^{-1}$	E_0/GPa	m	a
0	17.29	0.48	3.64	5.75	1.36
0-7	11.59	0.62	1.89	1.20	0.72
1-7	10.39	0.55	1.90	1.22	0.64
2-7	9.73	0.57	1.70	1.07	0.61
3-7	8.45	0.64	1.33	0.85	0.52
3-3	13.06	0.65	2.01	1.30	0.80
3-1	13.50	0.54	2.52	1.85	0.75

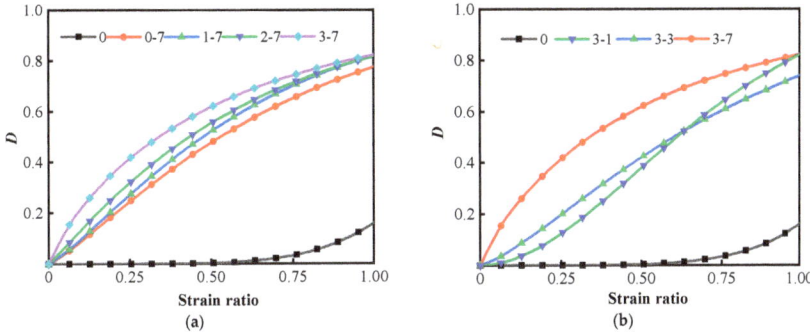

Figure 7. Evolution of some concrete damage, including (**a**) $N = 7$ and (**b**) $\lambda_c = 35\%$.

As shown in Figure 7b, at a stress level λ_c of 35%, the damage curve of the first wet–dry cycle concrete was first below the third and then rose above the third wet–dry cycle. The analysis showed that in the concrete curing process, there are some silicate components without a hydration reaction. Therefore, in the process of the wet–dry cycle, the internal joint weakness of the concrete will be destroyed by water erosion. However, the strong joint cannot be eroded by water, and thus the hydration reaction was carried out to strengthen the particles' association with each other. For the first and third wet–dry cycle damage curves, the early stage of the damage curve depended on the damage at the weak internal concrete joint [26,27], while the later stage depended on the hydration reaction at the strong internal concrete joint. As the number of wet–dry cycles increased, the damage variable D gradually increased from slow to rapid in the initial stage. This indicates that the concrete was more deeply affected by the deterioration. Summarizing the relevant literature [28,29] and as described in Figure 7, under coupled action conditions, the number of wet–dry cycles determined the lower limit of concrete deterioration, while the upper limit of concrete deterioration depended on the stress level of the sustained load.

4. Analysis of the Deterioration Mechanism of Concrete

4.1. Analysis of Concrete Acoustic Emission Energy Characteristics

The stress–strain curves reflect the macroscopic damage evolution of concrete. In addition, the study of the accompanying AE test parameters further analyzed the detailed information of the concrete specimens at different stages from a microscopic point of view. Concrete in the uniaxial compression process went through the compression dense stage, elastic stage, plastic stage, and post-peak stage. The corresponding acoustic emission detection went through three periods of calm, rising, and fluctuating phases, as shown in Figure 8a. The three periods of acoustic emission of concrete without the action of coupling conditions could correspond well to the four stages of the uniaxial compression process. According to the literature [30], it is known that the "cracking stress" of concrete is around 3 MPa. The corresponding stress level is between 10% and 20%. To visually compare the degree of deterioration effect of different coupling conditions on concrete, the stresses, AE energy release rates, and cumulative AE energy versus time for concrete under the action of coupled conditions with different stress levels experiencing seven wet–dry cycles, and stress levels of 35% experiencing different numbers of wet–dry cycles, are given, respectively.

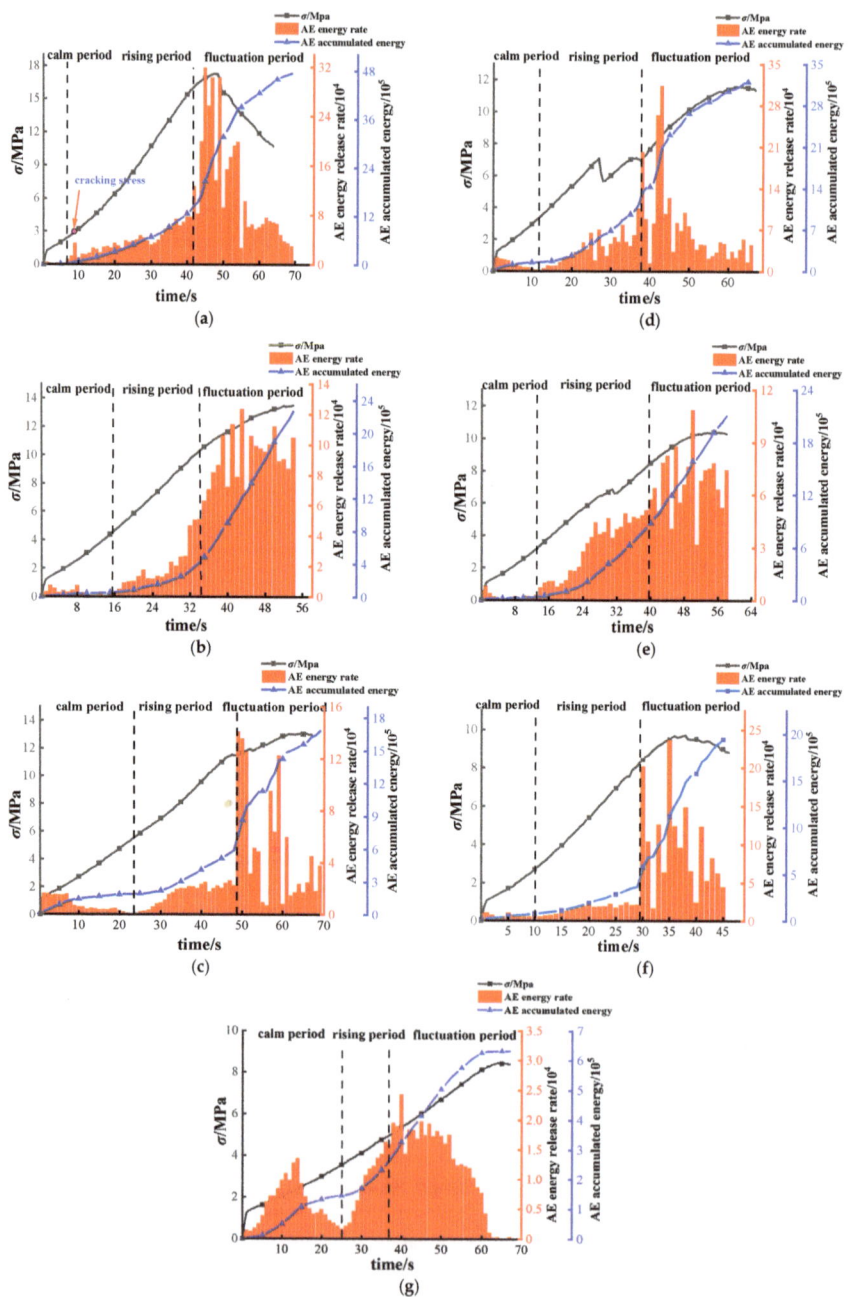

Figure 8. Concrete AE energy release rate, AE cumulative energy and stress versus time, including (a) $\lambda_c = 0\%$ $N = 0$, (b) $\lambda_c = 35\%$ $N = 1$, (c) $\lambda_c = 35\%$ $N = 3$, (d) $\lambda_c = 0\%$ $N = 7$, (e) $\lambda_c = 10\%$ $N = 7$, (f) $\lambda_c = 20\%$ $N = 7$, and (g) $\lambda_c = 35\%$ $N = 7$.

Comparing plots (b), (c), and (d) in Figure 8, the cumulative AE energy of the acoustic emission during the uniaxial compression test of concrete with increasing number of wet–dry cycles when the stress levels λ_c were all 35% decreased with increasing number of wet–dry cycles. The analysis suggested that the erosive action of water weakened or destroyed the concrete joint weaknesses. The drying process was then accompanied by the creation of secondary pores [29] within the concrete. Therefore, as the number of wet–dry cycles increased, secondary pore space was also created, thus increasing the extent of water erosion. These factors led to a continuous decrease in the concrete strength, an increasing growth rate of cumulative energy in the calm phase, an overall advance in the fluctuating phase, and a continuous advance in the maximum point of cumulative AE energy release rate. All of these phenomena indicate that the internal particles of concrete are more severely exploited by the wet–dry cyclic process, and the specimens gradually changed from brittle to plastic.

In the uniaxial compression process, the pores inside the concrete gradually closed with the increase in stress. When the stress reached the "cracking stress", the closed pores began to rupture and converge to develop cracks. Comparing (d), (e), (f) and (g) in Figure 8, the cumulative AE energy of the concrete decayed with the increasing stress level at seven wet–dry cycles. Plots (e) and (f) reflect the stages where the stress level was less than the cracking stress. Plots (d) and (g) reflect the stages where the stress level was greater than the cracking stress. A comparison of the two plots for each stress level phase shows that the cumulative AE energy share of both the calm phase and the rising phase of the specimen increased as the stress level increased. Although the laws are similar, the deterioration mechanisms are different. When the stress level is less than the initiating crack stress, the pore radius inside the concrete will keep decreasing. Because the depth of capillary action is inversely proportional to the radius of the capillary pores, the depth of water intrusion into the concrete specimen increases. This led to an increase in the dissolution effect of water on the micro-particles. The deterioration of concrete by this mechanism did not manifest itself in the first few wet–dry cycling processes, but was postponed over several wet–dry cycling processes. This was a result of the secondary cracking caused by each wet–dry cycling process. When the stress level approached the crack initiation stress, the fracture gradually closed to a size smaller than the capillary pore, thus reducing the water infiltration. Then, as the stress level as greater than the initiating crack stress [31,32], concrete cracks gradually developed and increased the contact surface of the water and micro particles. This deterioration mechanism on concrete often manifested itself during the first few wet–dry cycles. After several cycles, the crack surface was not spalled with micro particles. After this stage, the deterioration effect of wet–dry cycles on the concrete is not obvious. The above mechanism can be verified with the conclusions of Section 3.2.

4.2. Nuclear Magnetic Resonance T_2 Spectroscopy

Nuclear magnetic resonance (NMR) instrumentation utilizes the phenomenon of NMR generated by hydrogen nuclei in the presence of an applied magnetic field. The porosity of the specimen as well as the pore distribution were analyzed by measuring the transverse relaxation time (T_2) cutoff value of the saturated specimen. Among them, the size of the total peak area of the T_2 spectrum was related to the porosity of the specimen, and the position where the peak was located was related to the percentage of the pore size of the pore. The peak areas of each specimen after the action of different degradation parts are given in Table 6, and draw Figure 9, for example.

Combining Table 4 and Figure 9, the total peak area of the T_2 spectrum increased with the number of wet–dry cycles at stress levels λ_c of 0 and 10%. Where the stress level λ_c was 0, the peak area of the T_2 spectrum increased by 0.62% for the third wet–dry cycle compared to the first one. The seventh wet–dry cycle increased by 2.78% compared with the first one, and the overall pattern increased linearly. At a stress level λ_c of 10%, the T_2 spectral area increased by 1.50% in the third wet–dry cycle compared with the first. The seventh wet–dry cycle increased by 14.31% compared with the first one, with a parabolic progression in the overall pattern. The analysis concluded that when the stress level was less than the "cracking stress" stage, the depth of water infiltration into the concrete became deeper due to the capillary effect. Therefore, the dissolution effect on the micro-particles was more obvious, which led to a greater degree of degradation for a stress level of 10% than for a stress level of 0 in the wet–dry cycles. At stress levels λ_c of 20% and 35%, the T_2 spectral peak area decreased first and then increased with the increase in the wet–dry cycles. When the stress level λ_c was 20%, the T_2 spectrum area decreased by 2.25% after three wet–dry cycles compared with once, and increased by 0.62% after seven wet–dry cycles compared with once. At a stress level λ_c of 35%, the T_2 spectrum area decreased by 0.46% after three wet–dry cycles compared with once, and increased by 18.50% after seven wet–dry cycles compared with once.

Table 6. Peak areas of the T_2 spectra of concrete under different conditions.

Number	Total Peak Area/10^5	Proportion/%		
		First Peak	Second Peak	Third Peak
0	1.87	99.54	0.46	0
0-1	1.95	97.39	2.35	0.26
0-3	1.96	99.68	0.32	0
0-7	2.00	99.64	0.36	0
1-1	2.19	99.62	0.38	0
1-3	2.22	99.66	0.34	0
1-7	2.50	99.52	0.48	0
2-1	2.01	99.75	0.25	0
2-3	1.96	99.72	0.28	0
2-7	2.02	97.30	2.35	0.35
3-1	2.14	99.72	0.28	0
3-3	2.13	99.68	0.32	0
3-7	2.54	96.38	3.36	0.26

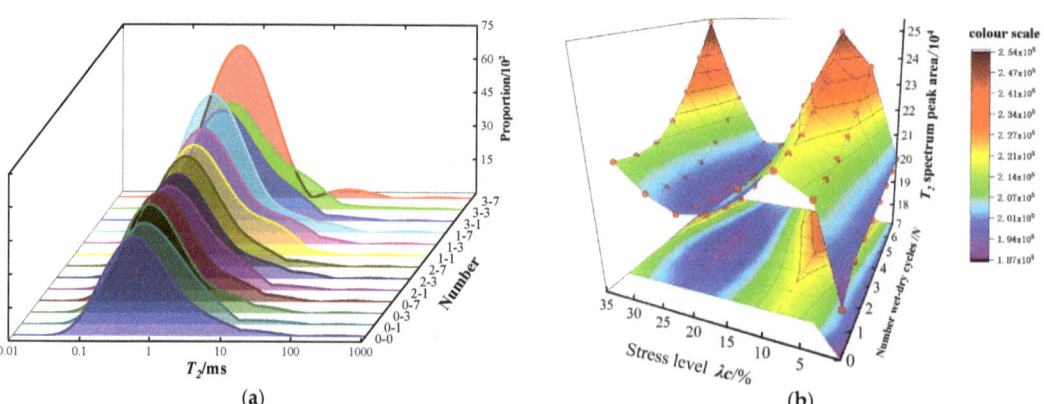

Figure 9. T_2 spectra of the specimens under different conditions, including the (a) T_2 spectrum distribution curve and (b) T_2 spectrum peak area.

After the analysis, the regenerative pore generation inside the concrete is shown in Figure 10. Under the condition of continuous loading, the internal edge pores were subjected to certain stress concentration. At this time and then after the impact of the wet–dry cycles on the edge pores, the strength here was reduced, and then under the action of continuous loading, the fissures were continuously expanded. There was an increase the contact area between the wet–dry cycle and the concrete internal particles. At the same time, the pores in the concrete developed and converge into cracks when the stress level was greater than the "crack initiation stress" stage. However, during the drying process, the water left the concrete in the gas and liquid phase [33], and some of the dissolved material was left in the cracks, which led to the phenomenon that the T_2 peak area of concrete at this stress level first decreased and then increased.

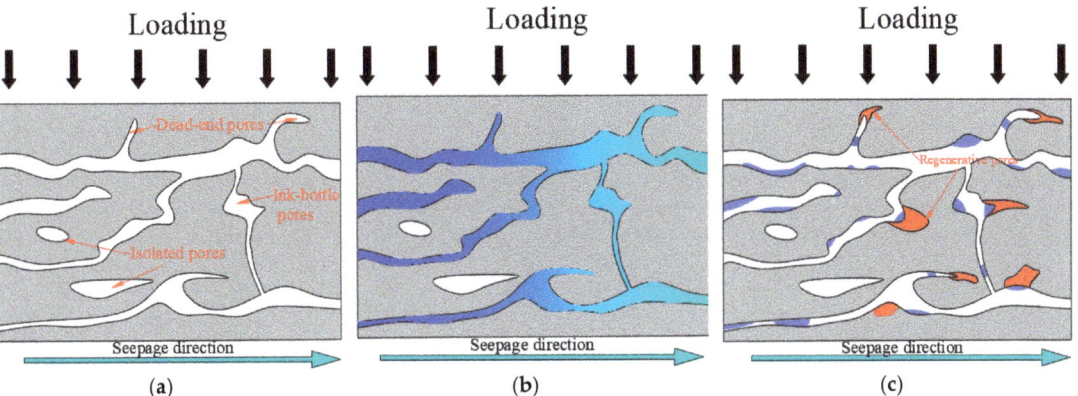

Figure 10. Concrete pore regeneration process: (**a**) initial diagram, (**b**) saturation diagram, and (**c**) naturally drying diagram.

5. Conclusions

(1) For the same number of wet–dry cycles, there are two thresholds for the effect of stress level on the uniaxial compressive strength of concrete a and b. The uniaxial compressive strength of concrete decreases with increasing the stress level in the interval from 0% to a. The stress level increases with increasing the load. In the interval from a to b, the concrete uniaxial compressive strength increases with the increase in load. In the interval b to 35%, the concrete uniaxial compressive strength again decreases with the increasing stress level. Meanwhile, the length of the interval from a to b decreases until it becomes zero, as the number of wet–dry cycles increases.

(2) The RationalTaylor regression model is used to better describe the variation of the uniaxial compression strength of concrete under the coupling conditions of different stress levels and the number of wet–dry cycles.

(3) Concrete AE evolution can be divided into three phases: calm phase, rising phase, and fluctuating phase. In the process of wet–dry cycle progression, the cumulative AE energy release percentage in the concrete calm phase stage increases continuously, and the fluctuation phase stage will gradually advance. Taking the crack initiation stress as the threshold, the concrete calm phase as well as the fluctuation phase change with an increment of the stress level in the phase greater than or less than the crack initiation stress, and the same law as for the wet–dry cycle progression.

(4) When the stress level is less than the cracking stress, the T_2 peak area of concrete increases with the progression of wet–dry cycles. However, when the stress level is greater than the cracking stress, the T_2 peak area decreases and then increases with the progression of the wet–dry cycles.

Author Contributions: L.W. and M.G. conceived and designed the theoretical framework; L.W. performed the experiment and collected the data; L.W. and J.Z. performed co-writing and revising of the manuscripts. All authors have read and agreed to the published version of the manuscript.

Funding: This work was supported by Innovation Fund for Graduate Students of Anhui University of Science and Technology (2021CX2015).

Institutional Review Board Statement: Not applicable.

Informed Consent Statement: Not applicable.

Data Availability Statement: The data presented in this study are available upon request from the corresponding author.

Acknowledgments: The authors would like to thank the project (2021CX2015) of the Anhui University of Science and Technology Graduate Innovation Fund.

Conflicts of Interest: The authors declare that they have no known competing financial interest or personal relationship that could have appeared to influence the work reported in this paper.

References

1. Tong, Y.Y.; Ye, L.; Ma, C. Mechanism Investigations of Realkalization for Carbonated Reinforced Concrete Based on Real Time Raman Spectroscopy Analysis. *J. Build. Mater.* **2017**, *20*, 894–901. (In Chinese) [CrossRef]
2. Jiang, C.; Gu, X.L. Discussion of "Assessing concrete carbonation resistance through air permeability measurements" by R. Neves et al. [Construction and Building Materials 82(2015): 304–309]. *Constr. Build. Mater.* **2016**, *102*, 913915. [CrossRef]
3. Ramesh, B.A.; Kondraivendhan, B. Effect of Accelerated Carbonation on the Performance of Concrete Containing Natural Zeolite. *J. Mater. Civ. Eng.* **2020**, *32*, 04020037. [CrossRef]
4. Jiang, W.Q.; Liu, Q.F. Chloride Transport in Concrete Subjected to Freeze-Thaw Cycles-A Short Review. *J. Chin. Ceram. Soc.* **2020**, *48*, 258–272. (In Chinese) [CrossRef]
5. Cao, D.F.; Ge, W.J.; Wang, B.Y.; Tu, Y.M. Study on the flexural behaviors of RC beams after freeze-thaw cycles. *Int. J. Civ. Eng.* **2015**, *13*, 92–101. (In Chinese) [CrossRef]
6. Zhang, S.; Zhao, B. Research on the performance of concrete materials under the condition of freeze-thaw cycles. *Eur. J. Environ. Civ. Eng.* **2013**, *17*, 860–871. [CrossRef]
7. Li, D.W.; Li, L.Y.; Wang, X.F. Chloride diffusion model for concrete in marine environment with considering binding effect. *Mar. Struct.* **2019**, *66*, 44–51. [CrossRef]
8. Yang, C.C. On the relationship between pore structure and chloride diffusivity from accelerated chloride migration test in cement-based materials. *Cem. Concr. Res.* **2006**, *36*, 1304–1311. [CrossRef]
9. Song, H.W.; Lee, C.H.; Ann, K.Y. Factors influencing chloride transport in concrete structures exposed to marine environments. *Cem. Concr. Compos.* **2008**, *30*, 113–121. [CrossRef]
10. Chen, J.; Jiang, M.; Zhu, J. Damage evolution in cement mortar due to erosion of sulphate. *Corros. Sci.* **2008**, *50*, 2478. [CrossRef]
11. Liu, K.W.; Cheng, W.W.; Sun, D.S.; Wang, A.G.; Zhang, G.Z. Effects of pH Value of Sulfate Solution on Calcium Leaching and Products of Cement Mortars. *J. Build. Mater.* **2019**, *22*, 179–185. (In Chinese) [CrossRef]
12. Zhang, J.; Yang, K.; He, X.; Wei, Z.; Zhao, X.; Fang, J. Experimental Study on Strength Development and Engineering Performance of Coal-Based Solid Waste Paste Filling Material. *Metals* **2022**, *12*, 1155. [CrossRef]
13. Huang, W.; Guo, Y.; Shen, A.; Li, Y.; Song, P. Numerical simulation of internal stress in pavement concrete under rolling fatigue load. *Int. J. Pavement Eng.* **2022**, *23*, 1306–1315. [CrossRef]
14. Wang, J.; Su, H.; Du, J.S. Influence of coupled effects between flexural tensile stress and carbonation time on the carbonation depth of concrete. *Constr. Build. Mater.* **2018**, *190*, 439–451. [CrossRef]
15. Wang, C.; Wu, H.; Li, C. Hysteresis and damping properties of steel and polypropylene fiber reinforced recycled aggregate concrete under uniaxial low-cycle loadings. *Constr. Build. Mater.* **2022**, *319*, 126191. [CrossRef]
16. Yu, H.F.; Sun, W.; Zhang, L.P.; Guo, L.P.; Li, M.D. Durability of concrete subjected to the combined actions of flexural stress, freeze-thaw cycles and bittern solutions. *J. Wuhan Univ. Technol.-Mater. Sci. Ed.* **2008**, *23*, 893–900. [CrossRef]
17. Su, X.P.; Wang, Q. Experiment of the Concrete Performance the Condition of Multiple Salts and Wet-dry Cycles. *J. Jilin Univ.* **2013**, *43*, 851–857. (In Chinese)
18. Wang, H.L.; Dong, Y.S.; Sun, X.Y.; Jin, W.L. Damage mechanism of concrete deteriorated by sulfate attack in wet-dry cycle environment. *J. Zhejiang Univ.* **2012**, *46*, 1255–1261. (In Chinese) [CrossRef]
19. Jiang, L.; Niu, D.T. Study of constitutive relation of concrete under sulfate attack and drying-wetting cycles. *J. China Univ. Min. Technol.* **2017**, *46*, 66–73. (In Chinese) [CrossRef]
20. Wang, S.Y.; Ding, L.; Hong, L. Durability study of high strength concrete beams strengthened with CFRP under sustained load and wet-dry cycle conditions. *Build. Struct.* **2017**, *47*, 78–83. (In Chinese) [CrossRef]
21. Gao, Y.F.; Ma, P.P.; Huang, W.P.; Li, X.B.; Cui, X.H. RRTS-II testing machine for rock rheological perturbation effect. *Chin. J. Rock Mech. Eng.* **2011**, *30*, 238–243. (In Chinese)
22. Cui, X.H.; Gao, Y.F.; Li, J.L. Research and Development of Rock Creep Experimental System under the Disturbing Load. *J. Shandong Univ. Sci. Technol.* **2006**, *25*, 36–38. (In Chinese) [CrossRef]
23. Zhou, C.Y.; Wei, J.X.; Yu, Q.J.; Yin, S.H.; Zhuang, Z.H.; Lei, Z.H. Water Absorption Characteristics of Autoclaved Aerated Concrete. *J. Wuhan Univ. Tech.* **2007**, *4*, 22–26. (In Chinese) [CrossRef]
24. Zeng, X.H.; Deng, D.H.; Xie, Y.J. Capillary Water Absorption Behavior of CA Mortar. *J. Southwest Jiaotong Univ.* **2011**, *46*, 211–216. (In Chinese) [CrossRef]

25. Wu, Z.; Zhang, C.J. Study of rock damage model and its mechanical properties under unidirectional load. *Chin. J. Rock Mech. Eng.* **1996**, 55–61. Available online: https://kns.cnki.net/kcms/detail/detail.aspx?FileName=YSLX601.007&DbName=CJFQ1996 (accessed on 14 September 2022).
26. Zuo, S.H.; Xiao, J.; Yuan, Q. Comparative study on the new-old mortar interface deterioration after wet-dry cycles and heat-cool cycles. *Constr. Build. Mater.* **2020**, *244*, 118374. [CrossRef]
27. Lv, Y.J.; Zhang, W.H.; Wu, F.; Li, H.; Zhang, Y.S.; Xu, G.D. Influence of Initial Damage Degree on the Degradation of Concrete Under Sulfate Attack and Wetting Drying Cycles. *Int. J. Concr. Struct. Mater.* **2020**, *14*, 47. [CrossRef]
28. Zhu, J.; Yu, R.; Han, S.; Tong, Y.; Zhang, H.; Zhen, Z. Strength Deterioration of Mudstone with Different Initial Dry Densities under Wet-dry Cycles. *J. China Railw. Soc.* **2021**, *43*, 109–117. (In Chinese)
29. Zhu, J.H.; Han, S.X.; Tong, Y.M.; Li, Y.; Yu, R.G.; Zhang, H.Y. Effect of Wet-dry Cycles on the Deterioration of Sandstone with Various Initial Dry Densities. *J. South China Univ. Technol.* (In Chinese). Available online: http://www.baidu.com/link?url=17 1VjgLuBjaYdBUERZWUq6sB_gSBpaaGH52tlPvGHlpPz49sgboULPF2Qy8zupm6cEKVC0mEDAecc9PkNuVpFYsSMiP-seE2 wxeTfdyjhmI9SDL6PhwgSbhpxOcBkqbr&wd=&eqid=ad740e3b0000415900000003634e33f6 (accessed on 14 September 2022).
30. Li, P.F.; Zhao, X.G.; Cai, M.F. Discussion on approaches to identifying cracking initiation stress of rocks under compression condition: A case study of Tianhu granodiorite in Xinjiang Autonomous Region. *Rock Soil Mech.* **2015**, *36*, 2323–2331. (In Chinese) [CrossRef]
31. Khanlari, G.; Abdilor, Y. Influence of wet–dry, freeze–thaw, and heat–cool cycles on the physical and mechanical properties of Upper Red sandstones in central Iran. *Bull. Eng. Geol. Environ.* **2015**, *74*, 1287–1300. [CrossRef]
32. Wang, L.Z.; Gao, M.Z.; Yang, D.C.; Wang, P. Capillary Water Absorption Properties of Concrete under Continuous Loading Coupled with Wet-Dry Cycle. *J. Bull. Chin. Ceram. Soc.* **2022**, *41*, 1998–2006. [CrossRef]
33. Ma, Z.; Shen, J.; Wang, C.; Wu, H. Characterization of sustainable mortar containing high-quality recycled manufactured sand crushed from recycled coarse aggregate. *Cem. Concr. Compos.* **2022**, *132*, 104629. [CrossRef]

Essay

Development and Constitutive Model of Fluid–Solid Coupling Similar Materials

Baiping Li *, Yunhai Cheng and Fenghui Li

School of Mining Engineering, Anhui University of Science and Technology, Huainan 232001, China
* Correspondence: libaiping0320@163.com; Tel.: +86-13013052181

Abstract: The Cretaceous Zhidan group (K1zh) pore fissure-confined water aquifer in Yingpanhao Coal Mine, Ordos City, China, has loose stratum structure, high porosity, strong permeability and water conductivity. In order to explore the fluid–solid coupling similar material and its constitutive model suitable for the aquifer, a kind of fluid–solid coupling similar material with low strength, strong permeability and no disintegration in water was developed by using 5~20 mm stone as aggregate and P.O32.5 Portland cement as binder. The controllable range of uniaxial compressive strength is 0.394~0.528 MPa, and the controllable range of elastic modulus is 342.22~400.24 MPa. The stress–strain curve and elastic modulus of similar materials are analyzed. It is found that the elastic modulus of similar materials with different water–cement ratios conforms to the linear law, the elastic modulus of similar materials with the same water–cement ratio after soaking treatment and without soaking treatment also conforms to the linear law. Based on the material failure obeying the maximum principal stress criterion and Weibull distribution, combined with the elastic modulus fitting formula, a constitutive model suitable for the fluid–solid coupling similar material was established, and the parameters of the constitutive model were determined by differential method. By comparing the theoretical stress–strain curve with the experimental curve, it is found that the constitutive model can better describe and characterize the fluid–solid coupling similar materials with different water–cement ratios and before and after soaking.

Keywords: water-bearing strata; fluid–solid coupling; similar materials; elastic modulus; linear law; constitutive model

1. Introduction

With the development of China's national economy, more and more deep underground projects are carried out in water conservancy, hydropower, transportation, energy development and national defense engineering [1–3]. For example, the Xiaolangdi Yellow River Diversion Project, Central Yunnan Water Diversion Project and Changdian Hydropower Station Diversion Tunnel Project have a higher buried depth. In global terms, the excavation of tunnels and underground space in China are characterized by large scale, large quantity and complex geological conditions and structural forms. In order to meet the needs of infrastructure construction, there are also more and more tunnels with large burial depth and large span and special length, which are often located in areas with strong geological structure. Complex geological conditions in deep underground engineering often lead to collapse, roof fall, water inrush and other disasters, resulting in significant economic losses and casualties [4–7].

Experts and scholars from various countries have conducted a lot of research on the prevention and control of underground engineering disasters and have achieved fruitful research results. Aissa Bensmaine [8], using the numerical code FLAC, carried out several numerical simulations in axisymmetric groundwater flow conditions to analyze the seepage failure modes of cohesionless sandy soils within a cylindrical cofferdam. They also indicate the sensitivity of the seepage failure mode to internal soil friction, soil dilatancy, interface friction and cofferdam radius. Moreover, new terms are proposed for the seepage failure mode designations based on the 3D view of the downstream soil deformation. It provides a new idea for studying the seepage failure of soil in cofferdam. For Özcan Çakır [9], the field compilation of the resistivity data is assumed to be completed by the application of the multiple electrode pole–pole array. The actual resistivity assembled underneath the analyzed area is inverted by considering the apparent (measured) resistivity values. Unique forms such as ore body, cavity, sinkhole, melt, salt and fluid within the Earth may be examined by joint interpretation of electrical resistivities and seismic velocities. Dang Van Kien [10] produced a numerical analysis using finite element software conducted to investigate the stability of rock mass surrounding the underground cavern and the system of caverns. The whole stability of surrounding rock mass of underground caverns was evaluated by Rocscience-RS2 software. This provides a reliable way to analyze the stability of the caverns and the system of caverns and will also help in the design or optimization of subsequent support. Bo Li [11], according to historical water inflow observation data of typical coal mines, produced a mine inflow prediction model based on unbiased grey and Markov theory was established. This model can eliminate the inherent bias of the traditional model and the effects of the random fluctuation of data on prediction results and has higher computational accuracy. The relevant research results can provide some basis for the improvement of the mine inflow method. Bo Li [12], according to the hydrogeological characteristics of southwest mines in China, produced twelve evaluation indicators determined from the three aspects of aquifer, aquifuge and geological structure, and an evaluation index system of water inrush risk of the karst aquifer in the coal floor has been constructed. On this basis, further using GIS technology and entropy weight theory, a multi-source information evaluation method for the risk of water inrush from the coal floor was proposed, and the evaluation results were verified.

At present, there are three main methods for stability analysis and disaster prevention of underground engineering: theoretical analysis, numerical simulation and model test. However, the complexity of the physical and mechanical properties of underground rock mass makes it difficult to deal with nonlinear problems in theoretical analysis, and it is difficult to simulate the real process of underground disasters by numerical simulation. Therefore, the results obtained by these two methods usually cannot accurately reflect the situation of diagenetic geological bodies in actual engineering conditions. However, the fluid–solid coupling simulation test can qualitatively or quantitatively reflect the complex construction technology, load action mode and time effect in practical engineering, and can also control the whole process from elasticity to plasticity of engineering stress and finally to the failure mode after the ultimate load. The key factor for the success of the fluid–solid coupling simulation test is to select a reasonable and reliable fluid–solid coupling similar material. The aquifer fluid–solid coupling similar material applied to the fluid–solid coupling simulation test needs to meet the following conditions: strong water permeability; the degree of shrinkage and expansion of the material after meeting water must be small; the material needs to have low strength so as not to soften and disintegrate after encountering water to maintain high integrity. As the basis of fluid–solid coupling simulation tests, the development of similar materials has experienced a tortuous and long development process. In the mid-1960s, Barton studied and manufactured a similar model material of raw material mixture composed of gypsum, red dan sand, coarse aggregate and water; this similar material has one characteristic: low elastic modulus. Jacoby [13] used glycerol and other materials as similar materials for model tests to study the problem of mantle convection and concluded that the upper lithospheric plate highly organized large-scale circulation, while the lower lithospheric plate became unstable at the core-mantle boundary. Ren Mingyang [14] developed a new similar material reflecting the fluid–solid coupling effect with iron powder, barite powder and quartz sand as aggregates, white cement as the cementing agent and silicone oil as the regulator. Zhang Ning [15] made a new similar material by mixing cement, sand, rubber powder, water, water reducer, early strength antifreeze and waterproofing agent, The compressive strength of this material is 15~50 MPa, and the elastic modulus is 2~4.5 Gpa. Zhang Jie [16] solved the problem of solid–liquid two-phase model material collapse in water by using paraffin as cementing agent in the study of fluid–solid coupling similar material. Li Shucai [17] developed similar materials (SCVO), which greatly improved the simulation of similar tests. In order to carry out the model test of the formation of high water pressure floor water inrush channel in deep mining. Sun Wenbin [18] developed similar materials suitable for deep well conditions. Li Shuchen [19] used paraffin wax as a cementing agent to develop a fluid–solid coupling similar material used in tunnel water inrush model test. Chen Juntao [20] developed a fluid–solid coupling similar material for the deep water resisting layer, realizing the similar simulation of the deep floor water resisting layer. Han Tao [21] developed a similar material to simulate porous rock mass and successfully applied it to the coupled model test of porous rock mass and the borehole wall. Dai Shuhong [22] used the orthogonal test design method to study fluid–solid coupling similar materials that can meet the similar physical and mechanical properties and similar water physical properties with talcum powder, gypsum and liquid paraffin as raw materials. Li Zheng [23] used clay, fine sand and glass fiber as raw materials to develop similar materials for surrounding rock, and used cement and carbon slag to prepare materials similar to the grouting environment, using multi-layer woven geotextiles to simulate a lining. Similar materials were successfully applied to tunnel seepage model tests. Wu Baoyang [24] taking quartz sand, barite powder and talc powder as aggregates, C325 white cement as the binder and silicone oil as the regulator, has developed a new type of fluid–solid coupling similar material with low strength, adjustable water absorption, non-disintegration in water and simple production. The feasibility and applicability of the material were verified by the model test of an underground reservoir in a multi coal seam mining coal mine.

In general, the research on fluid–solid coupling similar materials has made a certain degree of progress, but most of the fluid–solid coupling similar materials are mainly based on the study of aquifuge materials, the research on similar materials of aquifers is still lacking and most of the research work on similar materials pays more attention to solving the problem of easy disintegration of materials in water and lacks research on material constitutive models and failure modes. Therefore, it is urgent to carry out the research on fluid–solid coupling similar materials, as well as the research on their mechanical properties, constitutive models, failure modes and properties after water encounter.

2. Development of Fluid–Solid Coupling Similar Materials

2.1. Raw Material Selection

Before determining the raw materials, in order to reduce the workload, based on the principle of material selection, pre-experiment, a preliminary study was carried out to determine the raw materials of similar materials according to the function of the expected material and the physical and mechanical according to the function of the expected material and the physical and mechanical indices.

(1) Aggregate

As the main raw material of similar materials, aggregate plays a skeleton and supporting role in similar materials. If no aggregate is added during the production of the specimen, the specimen will not be formed. Aggregates are usually divided into coarse aggregates and fine aggregates. Coarse aggregates refer to materials with a diameter greater than 5 mm—generally gravel and pebbles. Fine aggregate refers to the material with a diameter of 0.165 mm, such as river sand, ore sand, sea sand, valley sand and quartz sand. The selection of aggregate by domestic scholars is also roughly within the above range. According to the selection principle of similar materials, it is required that the water content is stable, the texture is hard and clean and the grade is reasonable. Based on the above selection principle of similar materials, the aggregate selected in this paper is stone with a particle size of 5~20 mm.

(2) Cementing materials

Cementing material refers to the material with a certain strength that changes its own properties after physical and chemical action and can be closely bonded with other materials. Cement, gypsum, lime, clay, paraffin, rosin, epoxy resin, polyamide, and asphalt and commonly used cementing materials. In the study of fluid–solid coupling similar materials, the cementing materials used by domestic scholars usually include rosin, cement, clay, paraffin, cement, stone blue, latex, etc. In this paper, with reference to the cementing materials used in the physical model tests carried out in the past, and considering the advantages and disadvantages of each cementing agent, P.O32.5 Portland cement was finally selected. On the one hand, considering that the cementing strength of cement is between weak cementing materials such as gypsum and strong cementing materials such as cement, it can better adjust the strength of similar materials; on the other hand, compared with rosin and latex, cement is easy to use and less affected by temperature.

(3) Water

When preparing most similar materials, it is necessary to mix similar materials with water. The role of water in the preparation of similar materials mainly has two aspects. On the one hand, many cementing materials need water (i.e., hydration) when setting and hardening; on the other hand, when preparing similar materials and making similar material models, water must be used to meet the requirements of the production process. The similar materials developed in this paper choose water immersion to verify the performance of similar materials after encountering water.

2.2. Determination of Water–Cement Ratio

Based on the similarity principle, the strength of similar material is about 0.39~0.53 MPa. Through the strength test of different water–cement ratios, the strength of the material with a water–cement ratio of 0.9:1 is about 0.528 MPa. The test shows that the strength of the material is about 0.50 MPa when the water–cement ratio is 1.0:1, the strength of the material is about 0.46 MPa when the water–cement ratio is 1.1:1 and the material with a water–cement ratio of 1.2:1 is about 0.394 MPa. Therefore, it is finally determined to make specimens in the range of water–cement ratio 0.9:1~1.2:1, and measure the mechanical properties of different specimens.

2.3. Formulation of Similar Materials

In order to test the performance of similar materials with different ratios, six specimens were made in each ratio in the range of water–cement ratio of 0.9:1~1.2:1, and three specimens were in one group. The two groups of specimens were cured at room temperature standard curing conditions for 28 days. One group was directly tested for strength after curing, and the other group was soaked in water for 3~4 days and then tested for strength to test the performance of the specimen under water conditions. The size of the specimen was 150 mm × 150 mm × 150 mm. The test specimens were made as follows.

(1) Prepare a specific mold according to the design requirements, brush the mold with release agent and wait for the release agent to dry;
(2) Aggregate screening;
(3) The water and cement are weighed according to the design ratio, and the weighed water and cement are fully stirred for 5~10 min to make cement slurry;
(4) The sieved aggregate is poured into a uniformly stirred cement slurry and stirred for 5~10 min to ensure that the aggregate surface is uniformly hung with slurry;
(5) Put the stirred material into the mold;
(6) After standing for 48 h, the demolding was carried out, the label was affixed and the standard curing condition was maintained for 28 d.

The preparation process of similar materials is shown in Figure 1.

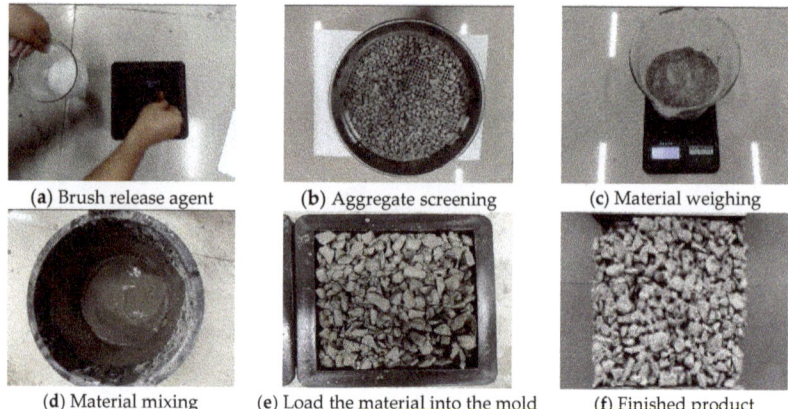

Figure 1. Preparation process of fluid–solid coupling similar materials.

3. Mechanical Properties Test and Analysis of Similar Materials

The uniaxial compression test of fluid–solid coupling similar materials was carried out on the RMT-150B rock mechanics test system. The axial load was used as the control index to load. The two groups of specimens were loaded until failure. The loading speed was 0.01 kN/s, and the stress–strain curve was recorded.

3.1. Mechanical Property Test of Specimens Not Soaked in Water

Uniaxial compression test is carried out on the test piece not soaked in water, and the stress–strain curve is obtained as shown in Figure 2:

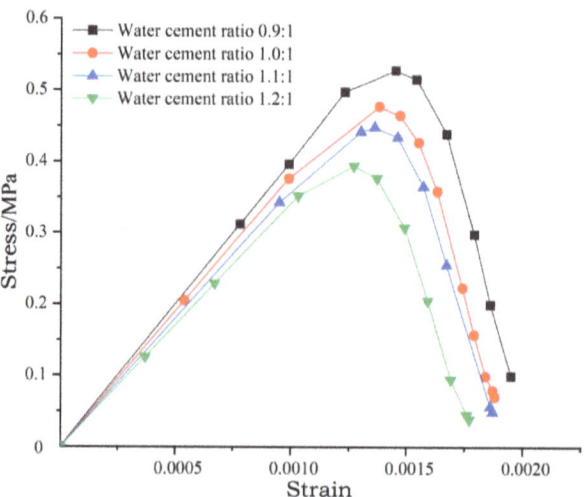

Figure 2. The stress–strain curve of the specimen without water immersion.

It can be seen from Figure 2 that the uniaxial compressive strength of the intact specimen without water immersion is 0.394~0.528 MPa, and the elastic modulus is 342.22~400.24 MPa. The elastic modulus is relatively stable when it does not reach the peak. The stress–strain trend of similar materials before the peak is approximately linear, indicating that the overall material is dominated by elastic deformation. After the peak, the stress–strain curve of similar materials begins to show more obvious plastic characteristics, indicating that there are new cracks inside the material. Under the action of uniaxial compressive stress, the evolution trend of the axial stress–strain curve of the specimen is similar, which is different from the compression curve of conventional rock materials. The fluid–solid coupling similar material used in this test is mainly elastic deformation before failure, so the stress–strain curve before the peak value shows an approximate linear evolution trend. After the stress passes the end of the elastic stage, it enters the post-peak failure stage. At this stage, the stress gradually decreases with the increase of strain, and there is an obvious nonlinear evolution trend. With the increase of water cement ratio, the peak strength of similar materials and the elastic modulus decrease gradually.

3.2. Mechanical Property Test of Specimens Soaked in Water

The uniaxial compression test is carried out on the specimens soaked in water, and the stress–strain curve is shown in Figure 3:

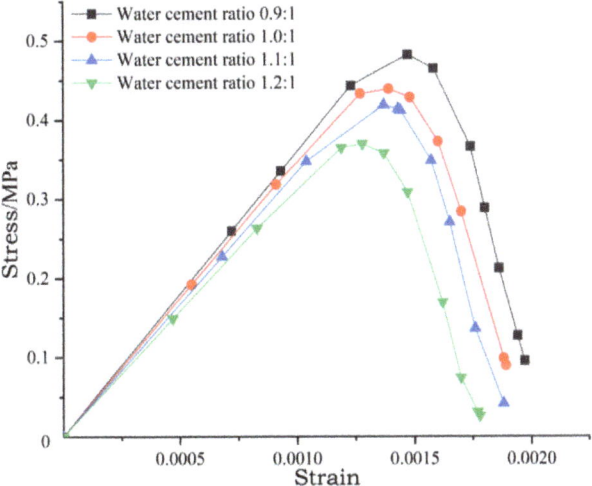

Figure 3. The stress–strain curve of the specimen soaked in water.

It can be seen from Figure 3 that the uniaxial compressive strength of the complete specimen after water immersion is 0.370~0.482 MPa, and the elastic modulus is 317.10~360.22 MPa. Compared with the material without water immersion, the peak strength and elastic deformation of the material after water immersion decrease to a certain extent, but the peak strain increases to a certain extent. Combined with previous studies on water chemistry [25,26], the material shows strain softening after water chemical erosion, and the failure mode changes from splitting to shear failure. In addition, after the rock material undergoes hydrochemical action, the particle skeleton reacts with the chemical composition to increase the internal pores. Therefore, under the same load, the displacement of the material soaked in water is larger than that of the material without soaking in water, the peak strain increases and the strength decreases.

3.3. The Peak Stress and Strain Evolution Law of Similar Materials with Different Water Cement Ratio

The evolution law of peak stress and strain of specimens with different water–cement ratios is analyzed. The peak stress and strain of specimens after immersion and those without immersion are shown in Figure 4.

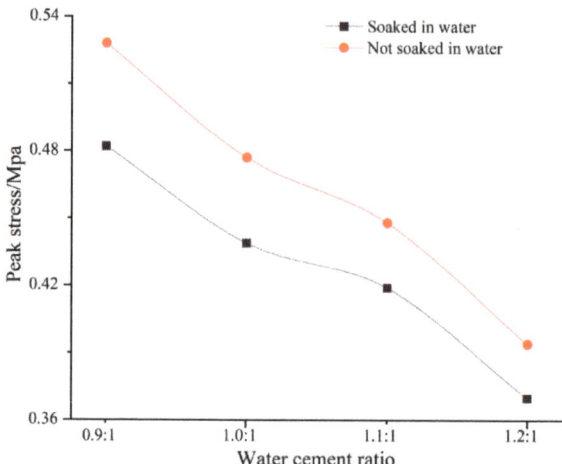

(a) The peak stress of specimens with different water cement ratio.

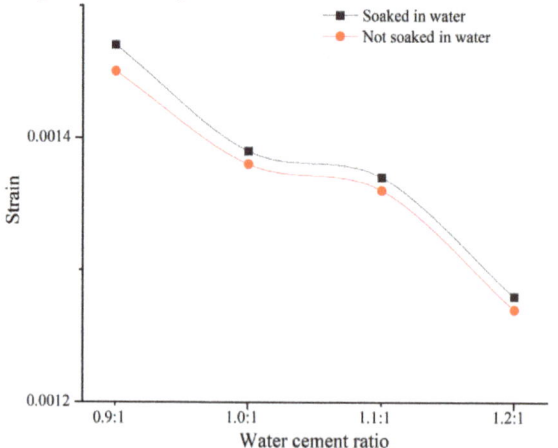

(b) The peak strain of specimens with different water cement ratio.

Figure 4. The evolution law of peak stress and strain of similar materials.

It can be seen from Figure 4 that the peak stress and strain of similar materials gradually decreases with the increase of water–cement ratio. In the same water–cement ratio specimen, the peak stress of the similar material soaked in water is significantly lower than that of the similar material not soaked in water, but the peak strain is slightly higher than that of the similar material not soaked in water. This is due to the chemical action of water, which leads to the reaction of the particle skeleton of the similar material with the chemical composition to increase the internal pores, so that the peak strain of the similar material soaked in water increases.

3.4. Evolution Law of Elastic Modulus of Similar Materials

In order to find the elastic modulus evolution law of similar materials with different water–cement ratios and similar materials before and after immersion, the elastic modulus of similar materials was studied. Figure 5a shows the evolution of the elastic modulus of similar materials with different water–cement ratios. Figure 5b shows the evolution of the elastic modulus of similar materials after soaking treatment and materials without soaking treatment. The results show that the elastic modulus is linear with the water–cement ratio and inversely proportional to the water–cement ratio. By analyzing the elastic modulus of fluid–solid coupling similar materials before and after soaking, it is found that the elastic modulus of the specimen after soaking is linearly related to the elastic modulus of the specimen without soaking.

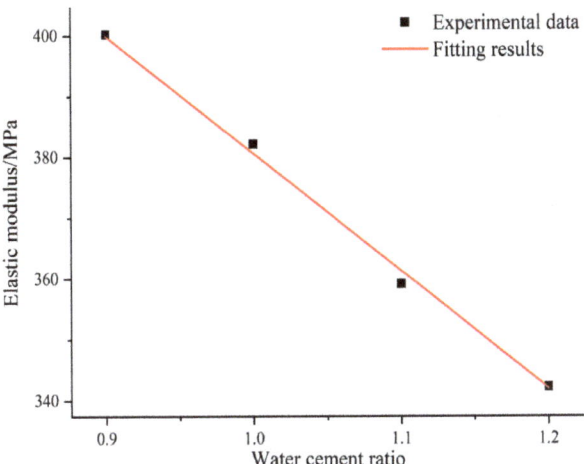

(a) The evolution law of elastic modulus of specimens with different water cement ratio.

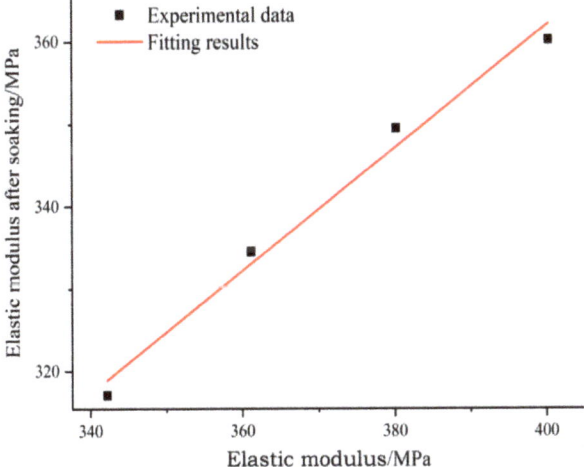

(b) Comparison of elastic modulus evolution of specimens before and after immersion.

Figure 5. Evolution law of elastic modulus.

By fitting the elastic modulus of similar materials with different water–cement ratios and before and after immersion, it is found that it conforms to the linear law, and an empirical formula for fluid–solid coupling similar materials is obtained.

$$E = -193.098\mu + 573.7044 \tag{1}$$

In the formula, E is the elastic modulus of fluid–solid coupling similar materials with different water–cement ratios, and μ is the water–cement ratio.

$$E_W = 0.74609E + 63.58443 \quad (2)$$

In the formula, E_W is the elastic modulus of fluid–solid coupling similar materials after immersion.

4. Constitutive Model

The stress–strain curves of fluid–solid coupling similar materials, especially the curves after the peak, are mainly nonlinear evolution characteristics, which cannot be accurately described and characterized by traditional elastoplastic models. The accurate description of stress–strain curve is of positive significance for understanding the stress evolution law of aquifer in similar model tests of fluid–solid coupling material and for further exploring the stress distribution of aquifer in practical engineering. Therefore, it is necessary to study the constitutive model of fluid–solid coupling similar material.

4.1. Weibull Distribution Damage Constitutive Model

In the mechanical analysis of materials, it is generally believed that there is a large amount of new crack initiation, crack propagation and old crack propagation in the plastic stage of the material; that is, the material has damage in the post-peak stage. In the equivalent strain hypothesis proposed by J. Lemaitre [27], it is shown that the effective stress is equal to the deformation of the damaged material, and the strain relationship of the damaged material can be expressed in the form of the non-destructive material. Replacing the nominal stress $[\sigma]$ with the effective stress $[\sigma^*]$, the damage constitutive equation of the post-peak stage of the material can be obtained as follows:

$$[\sigma] = [\sigma*](I - [D]) = [H][\varepsilon](I - [D]). \quad (3)$$

In the formula, $[\sigma]$ is the nominal stress, $[\sigma^*]$ is the effective stress, I is the unit matrix, $[D]$ is the damage variable matrix, $[H]$ is the elastic modulus matrix of similar materials, and $[\varepsilon]$ is the strain matrix.

Assuming that the damage of similar material is isotropic, the one-dimensional damage constitutive relation of similar material can be expressed as:

$$\sigma = \sigma^*(1 - D) = E\varepsilon(1 - D). \quad (4)$$

In the formula, D is the damage variable.

$$D = \frac{n}{N}. \quad (5)$$

In the formula, n is the number of destroyed elements, and N is the total number of elements of the lossless material.

According to the damage model method of Weibull distribution, the damage evolution law of material after peak load is explored. The probability density function is expressed as:

$$P(F) = \frac{m}{F_0}\left(\frac{F}{F_0}\right)^{m-1} \exp\left[-\left(\frac{F}{F_0}\right)^m\right]. \quad (6)$$

In the formula, $P(F)$ is the material element strength distribution function, F is the random distribution variable of element intensity, and m and F_0 are distribution parameters. The material damage variable can be defined as:

$$D = \int P(x)dx. \quad (7)$$

Therefore, the damage variable can be expressed as:

$$D = 1 - \exp\left[-\left(\frac{F}{F_0}\right)^m\right]. \tag{8}$$

Assuming that the micro-element strength obeys the maximum positive strain strength criterion, the damage variable can be expressed as:

$$D = 1 - \exp\left[-\left(\frac{\varepsilon}{\varepsilon_0}\right)^m\right]. \tag{9}$$

In the formula, D is the material damage factor, and m and ε_0 are the parameters related to the physical and mechanical properties of materials.

The material damage constitutive model can be described as:

$$\sigma = E\varepsilon \exp\left[-\left(\frac{\varepsilon}{\varepsilon_0}\right)^m\right]. \tag{10}$$

4.2. Determination of Constitutive Parameters

Formula (10) can be transformed into:

$$\frac{\sigma}{E\varepsilon} = \exp\left[-\left(\frac{\varepsilon}{\varepsilon_0}\right)^m\right]. \tag{11}$$

By taking logarithms on both sides of formula (11), we can obtain:

$$\ln\left(\frac{E\varepsilon}{\sigma}\right) = \left(\frac{\varepsilon}{\varepsilon_0}\right)^m. \tag{12}$$

In this paper, the uniaxial compression stress–strain curve of the fluid–solid coupling similar material is approximately linear before the peak, and there is an extreme point in the strain curve before and after the peak; that is, the slope of the stress–strain curve at the peak is 0:

$$\frac{d\sigma_c}{d\varepsilon_c} = 0. \tag{13}$$

In the formula: σ_c and ε_c represent peak stress and strain.

Substituting Formula (12) into Formula (13), we can get:

$$E \exp\left[-\left(\frac{\varepsilon_c}{\varepsilon_0}\right)^m\right]\left(1 - m\left(\frac{\varepsilon_c}{\varepsilon_0}\right)^m\right) = 0. \tag{14}$$

We arrange the formula (14) to obtain:

$$\left(\frac{\varepsilon_c}{\varepsilon_0}\right)^m = \frac{1}{m}. \tag{15}$$

We arrange the formula (15) to obtain:

$$m = \frac{1}{\ln\left(\frac{E\varepsilon_c}{\sigma_c}\right)}. \tag{16}$$

Substituting Formula (16) into Formula (15), we can get:

$$\varepsilon_0 = \varepsilon_c m^{\frac{1}{m}}. \tag{17}$$

Substituting Formulae (16) and (17) into Formula (10), we can get:

$$\sigma = E\varepsilon \exp\left[-\left(\frac{\varepsilon}{\varepsilon_c m^{\frac{1}{m}}}\right)^{\ln\left(\frac{E\varepsilon_c}{\sigma_c}\right)}\right]. \tag{18}$$

According to Formula (18) and (1), the constitutive model of fluid–solid coupling similar materials with different water cement ratio is:

$$\sigma = (-193.096\mu + 573.7044)\varepsilon \exp\left[-\left(\frac{\varepsilon}{\varepsilon_c m^{\frac{1}{m}}}\right)^{\ln\left(\frac{(-193.098\mu + 573.7044)\varepsilon_c}{\sigma_c}\right)}\right]. \tag{19}$$

According to Formulae (18) and (2), the constitutive model of fluid–solid coupling similar material after immersion is:

$$\sigma = (0.74609E + 63.58443)\varepsilon \exp\left[-\left(\frac{\varepsilon}{\varepsilon_c m^{\frac{1}{m}}}\right)^{\ln\left(\frac{(0.74609E + 63.58443)\varepsilon_c}{\sigma_c}\right)}\right]. \tag{20}$$

As shown in Figure 6, the stress–strain curve of similar materials with water–cement ratio of 1.1:1 is taken as an example to compare the experimental curve with the theoretical curve of the constitutive model. Taking Figure 6a as an example, in the stress stage of 0.155~0.355 MPa, the material deformation is elastic deformation, and the test curve is basically consistent with the theoretical curve. In the stage of stress 0.355~0.448 MPa, the test curve is slightly different from the theoretical curve, indicating that the material is not in complete elastic deformation before the peak, and there is a certain plastic deformation. After reaching the peak stress of 0.448 MPa, the material has obvious nonlinear deformation characteristics, and the experimental curve is approximately coincident with the theoretical curve. The results show that the constitutive model can better describe the stress–strain curve of fluid–solid coupling similar materials.

4.3. Failure form Analysis

Through the analysis of the model test results, it can qualitatively or quantitatively reflect the action mode and time effect of the load in the actual project and control the whole process of the elastic to plastic and the failure form after the ultimate load. Therefore, the study of the failure form of fluid–solid coupling similar materials is of positive significance for understanding the stress concentration and failure mechanism of aquifer in fluid–solid coupling model test.

Figures 7 and 8 show the morphology of similar materials with different water–cement ratios after failure. It can be seen from the figure that the larger the water–cement ratio, the worse the integrity of the material after failure. The reason for this phenomenon is that the larger the water–cement ratio, the smaller the cohesion between the particles of the material, and the looser the structure after failure. It can be seen from the figure that the specimens after immersion are the same as those without immersion. With the increase of water–cement ratio, the number of damaged blocks gradually increases. Comparing the test data and failure modes of the two specimens, the compressive strength and elastic modulus of the specimens after immersion were lower than those of the specimens without immersion. The structure of the specimens after immersion was looser than that of the specimens without immersion. This is due to the strain softening of the material after chemical erosion, and the failure mode is transformed from splitting to shear failure. In addition, after the rock material undergoes hydrochemical action, the particle skeleton reacts with the chemical composition to increase the internal pores.

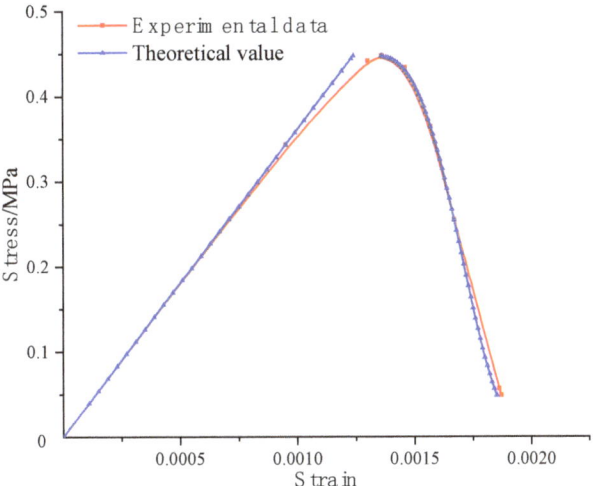

(a) Stress–strain comparison curve of specimens not immersed in water.

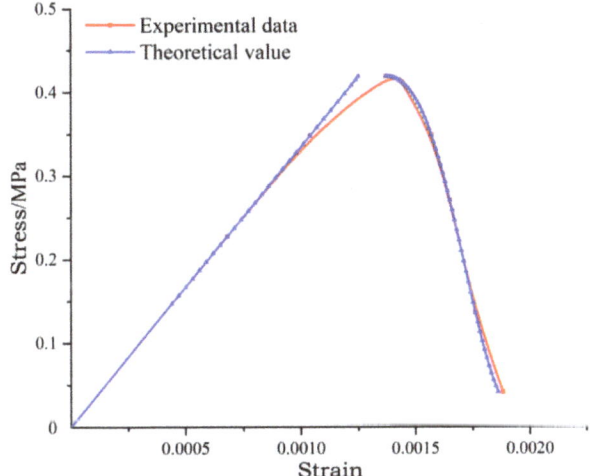

(b) Stress–strain comparison curve of specimens immersed in water.

Figure 6. Comparison between Theoretical Curve and Experimental Curve.

(a) 0.9:1   (b) 1.0:1   (c) 1.1:1   (d) 1.2:1

Figure 7. Failure form of specimens with different water–cement ratios not immersed in water.

(a) 0.9:1 (b) 1.0:1 (c) 1.1:1 (d) 1.2:1

Figure 8. Failure form of specimens with different water–cement ratios immersed in water.

4.4. Damage Analysis

It can be seen from Figure 9 that with the increase of strain, the damage–strain curves of similar materials show a nonlinear evolution trend. When the strain is equal, the greater the water cement ratio, the greater the damage. When the peak strain is reached, the damage occurs, and the damage growth rate is fast. With the increase of strain, the damage growth rate decreases and gradually becomes stable. When the material is close to failure, the damage growth rate of the material decreases again.

It can be seen from Figure 10 that with the increase of stress, the damage–stress curves of similar materials have nonlinear characteristics. When the stress is equal, the greater the water–cement ratio, the smaller the damage. The stress after the peak value of the material decreases with the increase of the damage. At the initial stage of the damage, the damage increases approximately linearly, and the growth rate is large. With the decrease of the stress, the damage growth rate also decreases, and gradually tends to be stable.

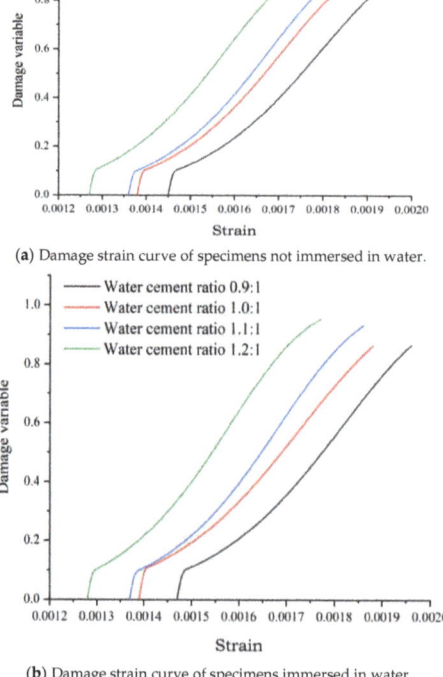

(a) Damage strain curve of specimens not immersed in water.

(b) Damage strain curve of specimens immersed in water.

Figure 9. Fluid–solid coupling similar material damage–strain curve.

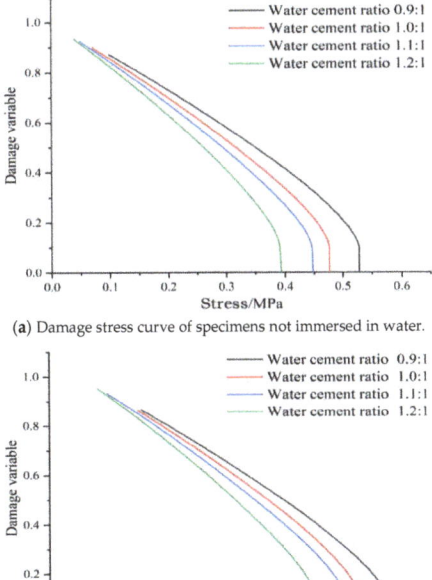

Figure 10. Fluid–solid coupling similar material damage–stress curve.

5. Conclusions

(1) A fluid–solid coupling similar material suitable for simulating the aquifer with loose formation structure, high porosity, strong permeability and water conductivity has been developed. The material has low strength, strong water permeability, and does not disintegrate when it meets water. The raw material is easy to obtain, and the production process is simple, which enriches the research means for aquifers with loose structure and strong water permeability.

(2) The peak stress and strain of similar materials gradually decrease with the increase of water–cement ratio; the elastic moduli of similar materials with different water–cement ratios follow the linear law, and the elastic moduli of similar materials before and after immersion also follow the linear law. Among the similar materials with the same water–cement ratio, the peak stress of the similar materials after immersion is significantly lower than that of the similar materials without immersion, but the peak strain is slightly higher.

(3) Based on Weibull distribution, the constitutive model of fluid–solid coupling similar material is established. The parameters of the constitutive model are further determined by the differential method. The constitutive model of fluid–solid coupling similar material is established by using the elastic modulus of similar material with a different water–cement ratio before and after immersion. Using Weibull distribution damage variable analysis, it is found that when the strain of fluid–solid coupling material is equal, the greater the water–cement ratio, the greater the damage caused by the material, and when the stress is equal, the greater the water–cement ratio, the smaller the damage caused by the material.

Although the constitutive model is obtained from the elastic modulus of fluid–solid coupling similar materials after immersion in water, there are still some deficiencies in the study of water rationality of similar materials, In this paper, the performance of similar materials is tested by immersing them in water, which verifies that they do not disintegrate when they meet water, but no specific permeability coefficient is given. In the next stage of research, we should focus on the water rationality of materials.

Author Contributions: Conceptualization, Y.C. and B.L.; methodology, Y.C.; software, B.L.; validation, Y.C., B.L. and F.L.; resources, Y.C.; data curation, Y.C. and B.L.; writing—original draft preparation, B.L.; writing—review and editing, Y.C. All authors have read and agreed to the published version of the manuscript.

Funding: This research received no external funding.

Institutional Review Board Statement: Not applicable.

Informed Consent Statement: Not applicable.

Data Availability Statement: Not applicable.

Conflicts of Interest: The authors declare no conflict of interest.

References

1. Zhang, W.; Zhang, D.S.; Shao, P.; Wang, X.F. Fast drilling and blasting construction technology for deep high stress rock roadway. *J. China Coal Soc.* **2011**, *36*, 43–48.
2. Ren, M.Y.; Zhang, Q.Y.; Chen, S.Y.; Yin, X.J.; Li, F.; Xiang, W.; Yu, G.Y. Experimental study on physical model of synergistic effect between lining and surrounding rock of large buried tunnel under complex geological conditions. *China Civ. Eng. J.* **2019**, *52*, 98–109.
3. Xie, H.P. Research progress of deep rock mechanics and mining theory. *J. China Coal Soc.* **2019**, *44*, 1283–1305.
4. Huang, Z.; Li, S.J.; Zhao, K.; Wu, R.; Zhong, W. Water inrush mechanism for slip instability of filled karst conduit in tunnels. *J. Cent. South Univ. (Sci. Technol.)* **2019**, *50*, 1119–1126.
5. Li, C.Y.; Wang, Y.C.; Liu, Y.; Jiao, Q.L.; Wang, M.T.; Zhang, Y. Model on variable weight–target approaching for risk assessment of water and mud inrush in intrusive contact tunnels. *J. Cent. South Univ. (Sci. Technol.)* **2019**, *50*, 2773–2782.
6. Yuan, Y.C.; Li, S.C.; Li, L.P.; Zhang, Q.Q.; Sun, B.L. Comprehensive analysis on disaster associated by water inrush and mudgushing in Shangjiawan Karst tunnel. *J. Cent. South Univ. (Sci. Technol.)* **2017**, *48*, 203–211.
7. Qian, Q.H. Challenges faced by underground projects construction safety and countermeasures. *Chin. J. Rock Mech. Eng.* **2012**, *31*, 1945–1956.
8. Bensmaine, A.; Benmebarek, N.; Bensmebarek, S. Numerical Analysis of Seepage Failure Modes of Sandy Soils within a Cylindrical Cofferdam. *Civ. Eng. J.* **2022**, *8*, 1388–1405. [CrossRef]
9. Çakır, Ö.; Coşkun, N. Theoretical issues with rayleigh surface waves and geoelectrical method used for the inversion of near surface geophysical structure. *J. Hum. Earth Future* **2021**, *2*, 183–199. [CrossRef]
10. Van Kien, D.; Ngoc Anh, D.; Ngoc Thai, D. Numerical simulation of the stability of rock mass around large underground cavern. *Civ. Eng. J.* **2022**, *8*, 81–91. [CrossRef]
11. Li, B.; Zhang, H.L.; Luo, Y.L.; Liu, L.; Li, T. Mine inflow prediction model based on unbiased Grey-Markov theory and its application. *Earth Sci. Inform.* **2022**, *15*, 855–862. [CrossRef]
12. Li, B.; Zhang, W.P.; Long, J.; Fan, J.; Chen, M.Y.; Li, T.; Liu, P. Multi-source information fusion technology for risk assessment of water inrush from coal floor karst aquifer. *Geomat. Nat. Hazards Risk* **2022**, *13*, 2086–2106. [CrossRef]
13. Jacoby, W.R.; Schmeling, H. Convection experiments and the driving mechanism. *Geol. Rundsch.* **2005**, *70*, 207–230. [CrossRef]
14. Ren, M.Y.; Yin, X.J.; Li, N.J.; Wu, X.Y.; Liu, H. Development and Application of Analogous Materials for Fluid-Solid Coupling Physical Model Test. *Adv. Mater. Sci. Eng.* **2022**, *2022*, 2779965. [CrossRef]
15. Zhang, N.; Li, S.C.; Li, M.T.; Yang, L. Development of a new rock similar material. *J. Shandong Univ. (Eng. Sci.)* **2009**, *39*, 149–154.
16. Zhang, J.; Hou, Z.J. Experimental study on seepage of coal mining under water. *J. Chengdu Univ. Technol. (Sci. Technol. Ed.)* **2009**, *36*, 67–70.
17. Li, S.C.; Zhou, Y.; Li, L.P.; Zhang, Q.; Song, S.G.; Li, J.L.; Wang, K.; Wang, Q.H. Development and application of a new similar material for underground engineering fluid-solid Coupled model testing. *Chin. J. Rock Mech. Eng.* **2012**, *31*, 1128–1137.
18. Sun, W.B.; Zhang, S.C.; Li, Y.Y.; Lu, C. Development application of solid-fluid coupling similar material for floor strata and simulation tests of water inrush in deep mining. *Chin. J. Rock Mech. Eng.* **2015**, *34*, 2665–2670.
19. Li, S.C.; Feng, X.D.; Li, S.C.; Li, L.P.; Li, G.Y. Research and development of a new similar material for solid-fluid coupling and its application. *Chin. J. Rock Mech. Eng.* **2010**, *29*, 281–288.
20. Chen, J.T.; Yin, L.M.; Sun, W.B.; Lu, C.; Zhang, S.C.; Sun, X.Z. Development and application of new solid-fluid coupling similar material of deep floor aquifuge. *Chin. J. Rock Mech. Eng.* **2015**, *34*, 3956–3964.
21. Han, T.; Yang, W.H.; Yang, Z.J.; Du, Z.B.; Wang, Y.; Xue, S.S. Development of similar material for porous medium solid-liquid coupling. *Rock Soil Mech.* **2011**, *32*, 1411–1417.
22. Dai, S.H.; Wang, H.R.; Han, R.J.; Wang, Z.W. Properties of similar materials used in fluid-solid coupling model test[J/OL]. *Rock and Soil Mech.* **2020**, *S2*, 1–8. [CrossRef]

23. Li, Z.; He, C.; Gao, X.; Yang, S.Z.; Luo, Y.W.; Yang, W.B. Development and application of a similar material for rock tunnel seepage model test. *J. Harbin Inst. Technol.* **2017**, *49*, 33–39.
24. Wu, B.Y.; Li, P.; Wang, Y.B.; Zhang, B.; Chi, M.B. Mechanical properties analysis of materials for similar simulation test of coal-water co-mining. *China Coal* **2022**, *48*, 64–73.
25. Zhang, X.W.; Xu, J.H.; Huang, N.; Sun, L.; Cao, Y. Mechanical properties and energy damage characteristics of sandstone subjected to hydrochemical erosion. *J. Min. Strat. Control Eng.* **2022**, *4*, 79–89.
26. Chen, Y.L.; Chen, Q.J.; Xiao, P.; Du, X.; Wang, S.R. A true triaxial creep constitutive model for rock considering hydrochemical damage[J/OL]. *Chin. J. Theor. Appl. Mech.* **2023**, *55*, 160–169.
27. Cao, W.G.; Zhang, S. Study on the statistical analysis of rock damage based on Mohr-Coulomb criterion. *J. Hunan Univ. (Nat. Sci.)* **2005**, *32*, 43–47.

Disclaimer/Publisher's Note: The statements, opinions and data contained in all publications are solely those of the individual author(s) and contributor(s) and not of MDPI and/or the editor(s). MDPI and/or the editor(s) disclaim responsibility for any injury to people or property resulting from any ideas, methods, instructions or products referred to in the content.

MDPI
St. Alban-Anlage 66
4052 Basel
Switzerland
www.mdpi.com

Sustainability Editorial Office
E-mail: sustainability@mdpi.com
www.mdpi.com/journal/sustainability

Disclaimer/Publisher's Note: The statements, opinions and data contained in all publications are solely those of the individual author(s) and contributor(s) and not of MDPI and/or the editor(s). MDPI and/or the editor(s) disclaim responsibility for any injury to people or property resulting from any ideas, methods, instructions or products referred to in the content.

www.ingramcontent.com/pod-product-compliance
Lightning Source LLC
LaVergne TN
LVHW070643100526
838202LV00013B/865